From Here to Infinity

THE ROYAL OBSERVATORY,
GREENWICH

FROM HERE
TO
INFINITY

A BEGINNER'S
GUIDE TO ASTRONOMY

John Gribbin & Mary Gribbin

STERLING

New York / London
www.sterlingpublishing.com

For Ellie, who asked lots of questions

Frontispiece: Spiral Galaxy M51 (NGC 5194), as seen in January 2005 by the Hubble Space Telescope. The galaxy is nicknamed 'the whirlpool'.

Library of Congress Cataloging-in-Publication Data Available

2   4   6   8   10   9   7   5   3   1

Published in 2008 by Sterling Publishing Co., Inc.
387 Park Avenue South, New York, NY 10016

First published in Great Britain in 2008 by the National Maritime Museum
Greenwich, London SE10 9NF
www.nmm.ac.uk/publishing

© John and Mary Gribbin, 2008

Distributed in Canada by Sterling Publishing
c/o Canadian Manda Group. 165 Dufferin Street
Toronto, Ontario, Canada M6K 3H6

For information about custom editions, special sales, premium
and corporate purchases, please contact Sterling Special Sales
Department at 800-805-5489 or specialsales@sterlingpub.com.

Manufactured in China

Designed by Clare Skeats
Diagrams by Greg Smye-Rumsby
Picture research by Sara Ayad
Copy-edited by Sarah Thorowgood
Project managed by Emily Winter
Production management by Geoff Barlow

Sterling ISBN 978-1-4027-6501-8

Contents

Introduction

A Brief History of Astronomy

How did we get here? How will it all end? Are there other civilisations yet to be discovered? Our fascination with such questions is a large part of what makes us human; it is hard to imagine even our cousins the gorillas wondering when the Universe was born. These questions used to be the province of religion and philosophy, but today they are the province of astronomy. What's more, they have answers – not perfect answers, but answers known to a precision that would have astonished anyone in previous centuries. The Universe around us emerged from a hot fireball, known as the Big Bang, 13.7 billion years ago. The Sun is one of hundreds of billions of stars in an 'island' called a galaxy, one of hundreds of billions of galaxies scattered across an expanding Universe. The Earth is a planet, one of a family of planets orbiting the much larger Sun. All the familiar chemical elements except hydrogen and helium were manufactured inside stars then recycled to make, among other things, later generations of stars, planets and, on at least one planet, people. You are literally made of stardust. The Sun and its family of planets are not unique; many other planetary systems are known within our own galaxy, so it is a reasonable inference that there are other civilisations on some other planets. And it will all end hundreds of billions of years from now, as the Universe expands faster and faster, carrying galaxies farther apart from one another as their last stars fade and die. But there is still a great deal to explore in the expanding Universe before that happens.

Astronomy today is one of the most exciting areas of scientific research. Astronomers study objects such as exploding stars and black holes, and they look back in time with their telescopes to probe the mystery of the origin of the Universe in the Big Bang. They search for planets orbit-

Fig. 1 (overleaf)
The 'Triangulum Galaxy' (M33). This 11 hr time-exposure mosaic is the most detailed ultraviolet image of an entire galaxy ever taken. It shows the giant star-forming region NGC 604 as a bright spot to the lower left of the galaxy's nucleus. With a diameter of 1500 light years (40 times that of the Orion Nebula), NGC 604 is the largest stellar nursery in the Local Group. Despite M33's small size, it has a much higher star-formation rate than either the Milky Way or Andromeda.

ing around other stars; hundreds of these have now been found, and it seems likely that soon we will find one similar to the Earth, our home planet. The life of an astronomer is often routine, even tedious, as they deal with computers and calculations to work out what the data flooding in from telescopes on Earth and in space are telling us about the Universe. But sometimes it is more glamorous and exciting, involving travel around the world to visit observatories high in the mountains of places like Chile (*fig. 2*) and Hawaii, or to attend meetings where new discoveries are announced and discussed. At one level, all this seems a far cry from the early days of astronomy; but in another sense, no discoveries made today can ever have the shattering impact of the first observations of the Universe made with the aid of a telescope.

The Birth of Astronomy

The scientific study of the heavens began almost exactly 400 years ago, in 1609, when the Italian Galileo Galilei (1564–1642; *fig. 3*) turned a telescope to the skies. He was not the first person to do this; but he was the first person to fully appreciate the significance of what he saw. And,

Fig. 2. An aerial view of Paranal Observatory in Chile, home of the European Southern Observatory, with VISTA, an infrared survey telescope, in the foreground and the Very Large Telescope in the background.

Fig. 3 (above) Galileo, by
Justus Sustermans, c.1639.

Fig. 4 (top right) Galileo's
telescope, 1609–10.

Fig. 5 (bottom right)
Galileo's drawings of the
movements of Jupiter's
moons in the Venice edition
of Sidereus Nuncius
(1610).

crucially, he told the world about his discoveries by writing a best-selling book, *Sidereus Nuncius* (*The Starry Messenger*), published in March 1610. It is a sign of how dramatic an impact Galileo's work had that within five years it had been translated into Chinese and published on the other side of the world. Galileo's observations destroyed the idea that the Earth is at the centre of the Universe, and that the Moon, Sun and other heavenly bodies are perfect objects supported by crystal spheres, orbiting it. The idea that the Earth orbits around the Sun had been around since the time of Nicolaus Copernicus (1473–1543), in the first half of the sixteenth century, but it was Galileo's observations that convinced many people of the truth of Copernicus' ideas.

Galileo destroyed the image of heavenly perfection by using his first telescope (fig. 4) to observe mountain ranges and craters scarring the surface of the Moon. This may seem trivial to us, but it was an important discovery 400 years ago; if the Moon can be seen to be imperfect, then the mergebat ceteris minor a Ioue tunc remota *m. 30.* sed paululû a recta linea versus Boream attollebatur, *vt apposita figura demonstrat.*

Die 27. hora 1. ab occasu, vnica tantum Stellula conspiciebatur, eaque orientalis secundum

Ori.   *   ☆   Occ.

hanc constitutionem: eratque admodum exigua, & a Ioue remota *min.7.*

Die 28. & 29. ob nubium interpositionem nihil obseruare licuit.

Die 30. hora prima noctis, tali pacto constituta spectabantur sydera: ynum aderat orientale,

Ori.   *   ☆   * * Occ.

a Ioue distans *min. 2. sec. 30.* duo vero ex occidente, quorum Ioui propinquius aberat ab eo *min. 3.* reliquum ab hoc *min.1.* extremorum & Iouis positus in eadem recta linea fuit, at media Stella paulum in Boream attollebatur: Occidentalior fuit reliquis minor.

Die vltima hora 2. visæ sunt orientales Stellæ duæ, vna vero occidua. Orientalium media a Io-

Ori.   * * ☆   *   Occ.

ue aberat *min. 2. sec. 20.* Orientalior vero ab ipsa media *min. 0. sec. 30.* Occidentalis distabat a Ioue *min. 10.* erant in eadem recta linea proxime, orientalis tantum Ioui vicinior modicum quiddam in Septentrione eleuabatur. Hora vero 4. duæ orientales viciniores ad inuicem

Ori.   *.   ☆   *   Occ.

adhuc erant; aberant enim solummodo *min. sec.*

20. apparuit in hisce obseruationibus occidentalis Stella satis exigua.

Die Februarii 1. hora noctis 2. consimilis fuit constitutio. Distabat orientalior Stella a Ioue

Ori.   ☆   *   Occ.

min.6. occidentalis vero 8. ex parte orientali Stella quædam admodum exigua a Ioue distabat *minutis secundis 20.* rectam ad vnguem designabant lineam.

Die 2. iuxta hunc ordinem visæ sunt Stellæ. Vna tantum orientalis a Ioue distabat min. 6. Iu-

Ori.   *   ☆   *   *   Occ.

piter ab occidentali viciniori aberat *min. 4.* inter hanc & occidentaliorem *min. 8.* fuit intercapedo; erant in eadem recta ad vnguem, & eiusdem fere magnitudinis. Sed hora septima, quatuor aderât Stellæ inter quas Iuppiter mediam occupabat se-

Ori.   *   *   ☆   *   Occ.

dem. Harum Stellarum orientalior distabat a sequenti *min. 4.* hæc a Ioue *min. 1. sec. 40.* Iupiter ab occidentali sibi viciniori aberat *m. 6.* hæc vero ab occidentaliori *min. 8.* erantque pariter omnes in eadê recta linea, secundum Zodiaci longitudinê extensâ.

Die 3. hora 7. in hac serie dispositæ fuerunt Stellæ. Orientalis a Ioue distabat *min.1. sec.30.* Occidentalis proxima *min. 2.* ab hac vero elongaba-

Ori.   *   ☆   *   *   Occ.

tur occidentalior altera *min. 10.* erant præcise in eadem recta, & magnitudinis æqualis.

1___These observations of Venus are particularly important in terms of demonstrating what makes a scientific theory scientific. A scientific idea that is offered to explain known observations, such as the Copernican idea that the planets orbit the Sun, is, strictly speaking, only a hypothesis until it is used to make predictions which can be tested and pass those tests. One of Galileo's former pupils, Benedetto Castelli (1578–1643), realised that Venus must show phases if it orbits the Sun and wrote to Galileo pointing this out. Galileo's observations confirmed the accuracy of the prediction. It was at that moment that the idea that the planets orbit around the Sun became a scientific theory, not just a hypothesis.

whole idea of regarding the Heavens as the province of God (or gods) and off-limits for scientific investigation is undermined. He showed that the planet Venus orbits around the Sun, by observing the changing phases of Venus, similar to the changing phases of the Moon as it orbits around the Earth (see page 43).[1] And he discovered four large moons orbiting around the planet Jupiter (*fig. 5*), an example in miniature of the way the planets, including the Earth, orbit around the Sun.

In effect, the work of Copernicus and Galileo removed Earth from the centre of the Universe in people's imagination. Removing the Earth from the centre of the Solar System was the first step in a long process, which continues today, that has shown with every new discovery that there is nothing special about our place in the cosmos. Once it became possible to measure distances across the Solar System, in the seventeenth century, it soon became clear that the Earth is far from being the largest or, except for being our home, the most important planet in the Solar System, and then later that the Sun itself is nothing special in the Universe. Galileo's telescopes (he made several) also showed that the band of light across the sky called the Milky Way since ancient times, is made up of a myriad of stars. In the seventeenth century, astronomers realised that the stars are objects like the Sun, but at such great distances that they appear only as pinpoints of light on the night sky. If the stars are suns, they also realised, then our Sun is just an ordinary star.

Beyond the Solar System

The Scot James Gregory (1638–1675) and England's Isaac Newton (1643–1727) tried to estimate the distances to the stars by comparing their brightness with the brightness of the Sun, and came up with distances for the nearest and brightest stars that were hundreds of thousands of times farther away than the Earth is from the Sun – once again reducing the perceived importance of our place in the Universe. And in 1750 the Durham astronomer Thomas Wright (1711–1786) published a book, *An Original Theory or New Hypothesis of the Universe*, in which he explained the appearance of the Milky Way on the sky by the idea that the Sun and all the stars we can see are part of a disc-shaped slab of stars that he described as being like the stone of a mill wheel (*fig. 6*). He even suggested that the Sun is not at the centre of the disc, but lies out to one side, in a place of no special importance within it. These ideas were taken up by the German philosopher Immanuel Kant (in 1755 Kant published the *Universal Natural History and Theory of the Heavens*), but it was only in the twentieth century that telescopes became powerful enough to test them properly and convert them from a hypothesis into a theory.

This is usually the way with science. It is no good having bright ideas unless they can be tested, and as we learn more and more about the Universe all the easy tests have been done and it gets harder and harder to test new ideas. Science and technology move forward together – science provides the basis for technology, and technology, such as the telescope, helps scientists to test their ideas. Until the twentieth century, just measuring the distances to the stars was a challenge. The first steps involve the same method used by surveyors here on Earth to measure distances. It is called triangulation (*fig. 8*). As long as you have a baseline

Fig. 6. *The heavens, according to Thomas Wright. Mezzotint (Plate XXXII) from his* An Original Theory or New Hypothesis of the Universe, Founded upon the Laws of Nature *(1750).*

2__Such as, the rule that the square of the length of the hypotenuse of a right-angle triangle is equal to the sum of the squares of the other two sides, and that the angles add up to 180 degrees.

with an accurately measured length, you can measure distances to objects without travelling to them. All you have to do is measure the angle of the line of sight to the object from each end of the baseline using a small telescope called a theodolite, then work out the distance from the rules of geometry applied to triangles.[2] If two observatories on opposite sides of the Atlantic Ocean measure the position of the Moon against the background of distant stars simultaneously, their observations can be used to measure the distance to the Moon, provided you know the distance between the two observatories (that involves other surveying techniques that we need not go into here). The distance to the planet Mars was measured in this way as long ago as 1671 by Giovanni Cassini and Jean Richer, of the Paris Observatory, and this led to one of the first accurate measurements of the distance to the Sun.

The planets orbit around the Sun in accordance with well-known laws that were discovered by Johannes Kepler (1571−1630) early in the seventeenth century and explained by Isaac Newton's law of gravity in the 1680s (fig. 7). This means that measuring distances to planets reveals the distance to the Sun. Distances to planets like Mars and Venus can now be measured very accurately by radar, and the distance to the Sun is known with equal accuracy to be 149,597,870 km (92,955,621 miles). This is also known as the 'astronomical unit' of distance. The diameter of the Earth's orbit around the Sun is just twice that, of course, and this gives us a baseline two astronomical units long that can be used in triangulation to measure the distances to the near-

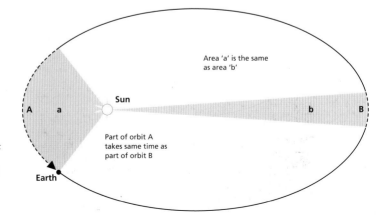

Fig. 7. The Earth's orbit around the Sun. This is slightly elliptical, as the Earth passes closer to the Sun at one end of the orbit than the other, causing it to move faster at that end of its orbit. As a result it 'sweeps out' equal areas in equal times.

est stars. The long baseline is used because of the vastness of the distances involved.

In astronomy, this triangulation technique is usually called parallax (*fig. 8*). It depends on seeing a nearby star seem to move against the background of distant stars as the Earth moves around its orbit, in just the same way that if you hold a finger up at arm's length and close each of your eyes in turn the finger seems to move against the distant background. Unfortunately, the technique only works for the nearest stars, because for most stars the distances involved are so great that no parallax can be observed. Even for the nearest stars, the change in angle measured in this way from one side of the Earth's orbit to the other is a few tenths of a second of arc. For comparison, the full Moon covers an angle of 30 minutes of arc on the sky. The angles being measured are about a 60th of one per cent of the angular size of the Moon. But in the nineteenth century astronomers did manage to measure the distances to a few stars in this way, and more distances were measured as the decades passed, better telescopes were built, and techniques such as photography became available.

The distances involved are so great that even astronomical units are too small to be practical for measuring them. The distance light can travel in one year, at a speed of 299,792 km (186,282 miles) per second, is 9.46 million million km (5.87 million million miles); this distance is called

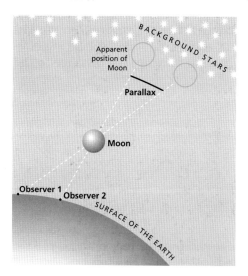

Fig. 8. *Triangulation. It is possible to calculate the distance of the Moon from the Earth by measuring its position against the background stars from two widely separated places at the same time.*

a light year. The nearest star to the Sun, Alpha Centauri, is 4.29 light years away. Light from Alpha Centauri takes 4.29 years to reach us – and even light from the Sun takes 8.3 minutes to reach us, so one astronomical unit is equal to 8.3 light minutes. Light years and light minutes are measurements of *distance* not of time. Thanks to satellite observations, astronomers now know accurate distances to thousands of relatively nearby stars, and are able to work out how the properties of stars, such as their colour and temperature, are related to one another. Armed with this information, they can work out distances to stars that are farther away, essentially by working out how bright a star must be from looking at its colour, then working out how far away it must be for a star that bright to look as faint as it does on the sky.

Early in the twentieth century, astronomers made a big breakthrough when they discovered that one family of stars behaves in a way that gives a more direct measurement of their brightness, and therefore their distance. These stars are called Cepheids, and they all vary in brightness in a con-sistent way, getting brighter and dimmer with a rhythm that repeats like a steady heartbeat for each individual star. But the 'heartbeat' is related to the average brightness of an individual star over its entire cycle. Brighter Cepheids take longer to go through this cycle (they have slower 'heart-beats') than fainter Cepheids, and measuring the period of the heartbeat of a Cepheid tells you how bright it is. Once again, this tells you how far away it must be to appear as faint as it seems from Earth. Many stars occur in groups, known as clusters, and if there is just one Cepheid in a clus-ter of stars it can be used to measure the distance to the whole cluster. One way and another, astronomers can now measure distances across the Universe quite accurately, and make maps of our surroundings. They can also tell what stars are made of, using the most important technique in the whole of astronomy – spectroscopy.

Secrets of Spectroscopy

A spectrum is the rainbow pattern of lines that is produced when white light passes through a prism. The colours are, in order of decreasing wavelength, red, orange, yellow, green, blue, indigo and violet. Isaac Newton was the first person to study in detail the way white light can be broken up into its constituent colours in this way, but it was not for another 200 years, until the middle of the nineteenth century, that astronomers discovered how useful spectroscopy could be. When they studied the rainbow pattern made by light from the Sun using a microscope, they found that the magnified spectrum is crossed by hundreds of dark lines, each one at a position corresponding to a precise wavelength of light. They also found the same kind of pattern of lines in light from stars. More than anything else, the patterns made by the lines resemble the patterns of the barcodes found on supermarket goods today; and like a barcode they tell you what the goods contain.

The key discovery, made by the German Robert Bunsen (1811–1899), whose name is forever associated with the Bunsen burner, is that each of the chemical elements produces its own unique, characteristic pattern of lines in a spectrum, as distinctive as a fingerprint. If light from a hot object is analysed in this way, the lines show up brightly in the spectrum; if white light is passed through a cold gas before being made into a spectrum, the lines show up as dark stripes. The dark lines in the spectrum of the Sun are

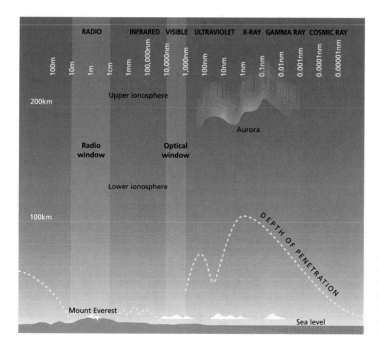

Fig. 9. *Electromagnetic radiation at different wavelengths is affected in different ways by the atmosphere of the Earth. Only radio waves and optical light penetrate easily to the ground.*

3__See John and Mary Gribbin, *Stardust: Supernovae and Life – The Cosmic Connection* (London: Allen Lane, 2000).

there because in the solar atmosphere there is gas that is cooler than the Sun itself, so it absorbs energy from below and makes the dark stripes in the spectrum. One particular set of lines reveals that there is hydrogen in this gas, another set tells you that there is iron, and so on. Even better, the intensity of each set of lines tells you how much of each element there is, relative to the other elements.

One of the most important things about this discovery was that nobody had to know how the elements made these patterns of lines. People like Bunsen could study the light from different elements in the laboratory and measure the positions of the lines, then look at sunlight (and starlight) to see if the same patterns were present there. The lines were only explained in the twentieth century, using quantum physics, but for astronomical purposes there is no need to discuss that here.[3]

One of the greatest early triumphs of spectroscopy came in 1868, when the astronomer Norman Lockyer (1836–1920) found a set of lines in the spectrum of light from the Sun that did not match any element known on Earth. He decided that there must be an element present in the Sun that had never been detected on Earth, and called it helium, from the Greek word for the Sun, *helios*. Helium was only found on Earth in 1895, by William Ramsay, and Lockyer was knighted for his work two years later.

The development of spectroscopy went hand in hand with another nineteenth century development: photography. With photographs, astronomers could get images of stars and other astronomical objects to study at their leisure. Photographs also store more light than the human eye, so they can make images of stars too faint to be seen with the naked eye, or even by the human eye aided by a telescope. And with the use of spectrophotography astronomers could take spectra back to their offices and laboratories to study and share with their colleagues, instead of straining their eyes to pick out the faint lines while in the dark of an observatory, bent over a microscope being fed light from a spectroscopic apparatus attached to a telescope.

Spectroscopy does something else as well. It tells us how fast stars are moving. If a star is moving towards us, the light from it gets squashed to shorter wavelengths and the whole pattern of lines shifts towards the blue end of the spectrum (it might seem more logical to describe this as a shift towards the violet end of the spectrum, but traditionally astronomers identify the short-wavelength end as blue). If the star is moving away, the light gets stretched, and the pattern shifts towards the red end of the spectrum. These blueshifts and redshifts are collectively known as Doppler shifts, or the Doppler effect, after the German Christian Doppler (1803–1853), who studied the equivalent processes in sound waves in the middle of the nineteenth century. They are exactly the same as the way the pitch from the siren of a fire truck racing towards you sounds higher than when the same siren is moving away from you.

By the end of the 1920s, astronomers were able to measure the distances to stars using the techniques we have described, including the Cepheid technique, and others. They knew what stars were made of, thanks to spectroscopy, and how fast they are moving, thanks to the Doppler effect. They were about to discover the true scale of the Universe. With improving technology, the first really large telescopes, built on mountain tops, were becoming available, and with photography they had the means to make permanent images of very faint objects and to study their spectra in detail and at length. Modern astronomy really began when all these observational techniques were linked with the understanding of the physics of atoms and particles – quantum physics – in the 1920s. That would explain how stars work, and where the chemical elements came from. Everything else, including observations from satellites orbiting above the obscuring layers of the Earth's atmosphere, radio telescopes, computers and digital imagery, has been built on that base, in less than a hundred years – less, indeed, than a human lifetime. In this book, we shall not go into all the details of how astronomers have learned what they know about the Universe, but will concentrate on the discoveries that they have made. These alone would be enough to fill several books; astronomers can tell the whole story from how the Universe as we know it started in a Big Bang 13.7 billion years ago and how stars, planets and people emerged to what will happen to the Sun and stars in the future, and even, with a little guesswork, what will be the fate of the entire Universe.

But it makes sense to start our tour of the Universe closer to home, here and now on Planet Earth, and work our way out into the Universe at large.

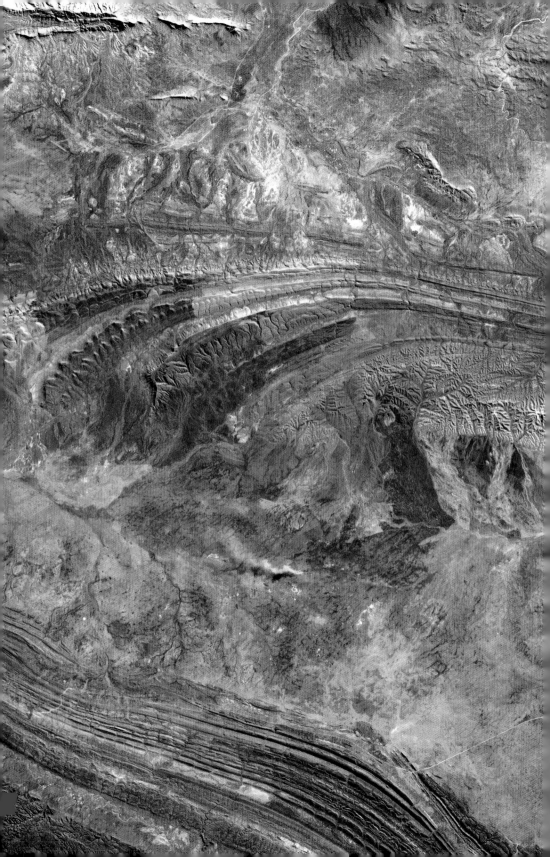

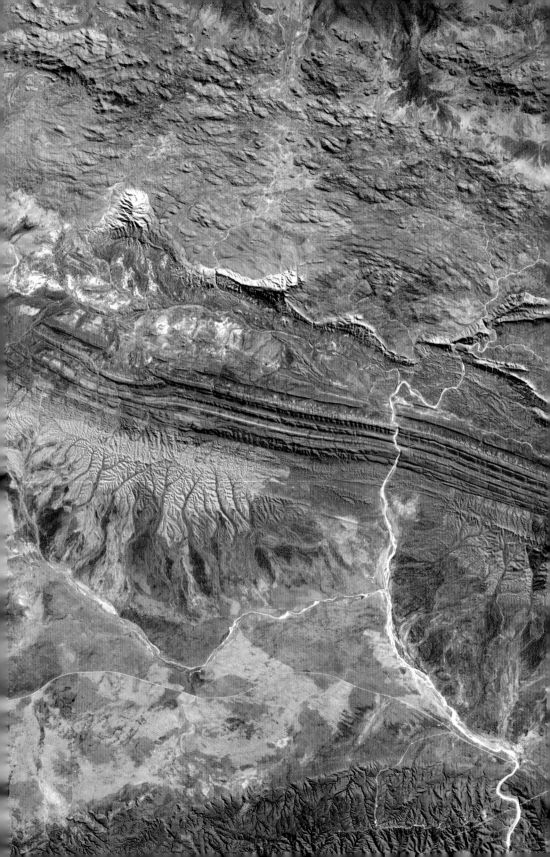

Chapter_One

Earth

We live on a ball of rock covered with a thin layer of water and gases, turning on its axis once every day and orbiting around the Sun once every year. Although it is special to us as our home in space, it is an ordinary planet, one of several in the Solar System. And it is very definitely not the centre of the Universe.

Images of the Earth from space are so familiar in the twenty-first century that to most people this description of our home in the Universe is so obvious that it hardly needs pointing out (fig. 11). Yet only four centuries ago, in Galileo's time, many people had trouble believing the truth even when observations made with telescopes showed that it must be the case. There were two reasons why this was so. First, people wondered why, if we are moving through space, we cannot feel the motion. In those days, travelling at high speed meant riding a galloping horse, and you certainly knew you were moving then, or if you were jolted around in a carriage being pulled by galloping horses. Today, we are used to travelling in aircraft at hundreds of kilometres an hour, moving smoothly through the air without noticing our speed, eating, drinking, walking about and doing many of the everyday things we do on the surface of the Earth. We understand as part of our own common sense that what matters is not how fast you are moving but how much your speed changes. We only notice the movement of the aircraft when it accelerates down the runway for takeoff, or when it decelerates on landing – or when it is jolted around by turbulence.

The second thing that puzzled our ancestors about the idea of a round Earth orbiting in space was why people on the other side of the world, at the 'bottom', did not fall off. It was only later in the seventeenth century that Isaac

Fig. 10 (overleaf)
One spectacular element of the Earth's surface is a crater in Australia's Northern Territory that is 24 km (15 miles) in diameter and 5 km (3 miles) deep, formed 142 million years ago by an asteroid or comet.

Fig. 11 (right) *This dramatic 'blue marble' image is the most detailed true-colour mosaic image of the Earth, using a collection of satellite-based observations of the land surface, oceans, sea ice and clouds.*

Fig. 12. *Life in the sea includes colonies of coral which appear vibrant blue-green in this image of Australia's Great Barrier Reef. Stretching 2000 km (1240 miles) along the northeast coast of Australia, the 'reef' is in fact made up of many individual detached reefs, separated by deep-water channels.*

Newton explained how the force of gravity pulls things towards the mathematical centre of an object like the Earth. Wherever you are on Earth, 'down' is always in the direction towards the centre of the Earth. A person standing at the North Pole may be 'upside down' from the point of view of a person standing at the South Pole, but they are both being tugged towards the centre of the Earth by gravity. Newton also explained how the same force keeps the Moon in orbit around the Earth, and the Earth in orbit around the Sun.

The Grip of Gravity

Newton realised that everything in the Universe will keep moving in a straight line at a steady speed unless it is pushed or pulled by a force. This is not obvious on the Earth, because here there is always a force, called friction, trying to slow things down. But Newton imagined how the Earth moves in space without any friction. It 'wants' to move forward in a straight line, a tangent to its orbit around the Sun, but every time it moves a tiny bit forward the Sun's gravity tugs it a tiny bit sideways. The combined effect of the two movements is that the Earth's path bends a little towards the Sun, and over a whole year all this bending brings the path round in a closed orbit, back to where it started. It is one thing to guess this might happen; but, famously, Newton's clever contribution was to prove mathematically that a force of gravity, which is proportional to 1 divided by the square of the distance between two objects, would explain the fall of an apple from a tree, and the Moon's orbit around the Earth. This means that if something is twice as far away it feels one-quarter of the force (because $2^2 = 4$) three times farther away it feels one-ninth of the force ($3^2 = 9$) and so on. The same inverse square law explains the orbit of the Earth around the Sun, and the orbits of other objects in the Universe, such as an artificial satellite orbiting round the Earth, or a spaceprobe on its way to Jupiter.

Your weight depends on how much stuff there is in your body (its mass), and on how hard the Earth pulls. The more massive an object is, the harder it pulls. An astronaut on the surface of the Moon has the same mass as on Earth, but because the Moon is much smaller than the Earth, it does not pull as strongly as the Earth. On the Moon, you would weigh only one-sixth as much as you do on Earth. Your mass is always the same, wherever you go in the Universe, but your weight depends on where you are. In space, an astronaut is literally weightless.

Gravity also explains why the thin layer of water and gases around the solid Earth is smeared out uniformly over its surface, like the skin of an apple. It is all being tugged towards the centre of the Earth. So the atmosphere also has weight, squashing the lower layers and producing a pressure at the surface of the Earth where we live. But because it is squashed in this way the atmosphere is very thin compared with the size of the Earth. The diameter of the Earth is 12,756 km (7926 miles). Because gravity is pulling everything towards the centre of the Earth, the density of the atmosphere quickly gets less farther above the surface. In this sense, the atmosphere gets thinner (less dense) as you climb a mountain. At the top of the tallest mountain on Earth, Mount Everest, the air is barely dense enough for people to breathe.

Mount Everest is just under 9 km (8.85 km, 5.5 miles) high. The distance from the bottom of the deepest part of the ocean to the height in the atmosphere equivalent to the top of Mount Everest is only about 20 km (12.4 miles), about 0.15 per cent of the diameter of the Earth. This layer represents the region of the Earth where life is possible. If the Earth were a ball the size of a grapefruit, the zone of life would be a layer just 0.2 mm thick over the surface of the ball: no more than a coat of paint.

The Air We Breathe

So space is much closer to us than most people realise. There are layers of the atmosphere above the breathable

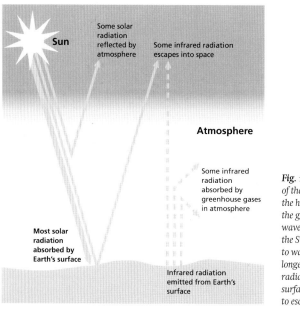

Sun

Some solar radiation reflected by atmosphere

Some infrared radiation escapes into space

Atmosphere

Some infrared radiation absorbed by greenhouse gases in atmosphere

Most solar radiation absorbed by Earth's surface

Infrared radiation emitted from Earth's surface

Fig. 13. The atmosphere of the Earth traps some of the heat from the Sun by the greenhouse effect. Short wavelength radiation from the Sun penetrates easily to warm the surface, but longer wavelength radiation from the warm surface finds it harder to escape back into space.

region near the ground, but even they do not extend far above the surface. The bottom layer of the atmosphere is called the troposphere, and extends up to about 15 km (9.3 miles). Temperature goes down as you go up through the troposphere, because the air traps heat near the ground. This is called the greenhouse effect (although actually it is not the way a greenhouse traps heat). What happens is that most sunlight passes through the atmosphere and warms the surface of the Earth. The warm surface radiates infrared heat, like the warmth you feel if you hold your hand near, but not touching, a radiator. Gases in the troposphere, especially carbon dioxide and water vapour, absorb infrared radiation from the surface of the Earth and warm up the bottom of the atmosphere (fig. 13).

Fig. 14. Cumulus clouds are visible in this photograph, taken in February 1984 by an astronaut onboard the Space Shuttle. It shows a series of mature thunderstorms located near the Parana River in southern Brazil.

Fig. 15. The Earth's atmosphere: clouds hanging above the cold Benguela current, which travels northward along the Atlantic coast of south-western Africa, as viewed by the Space Shuttle in 2002.

From about 15 km to 50 km (9.32 to 31.1 miles), in a layer called the stratosphere, although the air is very thin it warms up slightly, because the ozone in this layer absorbs some of the incoming energy from the Sun; this is sometimes called the ozone layer. Above 50 km (31.1 miles) altitude, the temperature falls away again. By the time you get to 100 km (62.1 miles) altitude the air is so thin that the idea of temperature in the everyday sense has no meaning. The first cosmonaut, Yuri Gagarin (1934–1968), orbited the Earth at an altitude between 190 km (118 miles) and 344 km (214 miles); at 200 km (124 miles) above the surface of the Earth, you are already in space.

This is a tiny distance, 1.5 per cent of the diameter of the Earth, and far less than the distances many people travel for business or for holidays. The difference is that starting from London and travelling 200 km (124 miles) across the surface of the Earth takes you as far as Devon or Belgium, but starting from London and travelling 200 km (124 miles) straight up takes you into space.

The thin smear of the life zone around the Earth is important to us, and gives the Earth a distinctive appearance from space, with the blue of the oceans contrasting with the white of the clouds. By analysing the light from the Earth using spectroscopy, any alien astronomers with good enough telescopes would also be able to work out what the atmosphere of the Earth is made of. They would discover that it is rich in oxygen, which is a highly reactive gas, and they would probably realise that this is an important clue as to what is going on at the surface of the Earth.

If there were no life on Earth, the oxygen in the air would soon react with other material to make stable gases such as carbon dioxide. Within a few million years, the oxygen would be gone. An atmosphere rich in non-reactive gases such as carbon dioxide is said to be in a stable equilibrium. An atmosphere rich in reactive gases like oxygen is unstable. It can only stay in the same state for a long time, far away from equilibrium, if oxygen is constantly being manufactured and released into the air to replace the oxygen that is being used up in chemical reactions. The atmosphere of the Earth is maintained in this unstable situation by the actions of life. At its simplest, plants use sunlight to break carbon dioxide down into carbon and oxygen, using the carbon to build the structure of their tissues and releasing oxygen into the air. Animals breathe in oxygen to use as a source of energy, combining it with carbon from their food in a form of slow burning and breathing out carbon dioxide. The overall picture is more complicated, but this gives a rough idea of what is going on.

Forty years ago, a British scientist working for NASA, James Lovelock (b. 1919), realised that this kind of process provides a test for life on other planets.[4] If planets have atmospheres in stable equilibrium, they cannot support life. If planets have unstable atmospheres rich in gases like oxygen, they must be homes for life. Over many years, from this insight he has developed the idea that all of the living things on Earth, *and* non-living components such as volcanoes spewing out gases, are part of a single system, in the same way that all of the cells in your body, and all the organs such as your heart, lungs and liver, and bones, are part of a single living thing. Lovelock calls the living Earth 'Gaia', after the name of the Greek goddess of the Earth (fig. 16). (Scientists who are uncomfortable with the idea of naming theories after Greek goddesses prefer to talk about Earth System Science, which also means studying the Earth as a

4__See James Lovelock, *Homage to Gaia: The Life of an Independent Scientist* (Oxford: Oxford University Press, 2000).

Fig. 16. *James Lovelock, the founder of Earth System Science, with a statue of Gaia.*

single system involving living things and non-living things interacting with one another.)

One aspect of Gaia/Earth System Science is that life controls the amount of carbon dioxide in the air. If plants are more active, they use up more carbon dioxide. If they are less active, more carbon dioxide is taken from the air. Because carbon dioxide contributes to the greenhouse effect, this means that life on Earth affects the temperature of the whole planet.

Whatever its name, this is one of the most important areas of scientific research today. At a practical level, it is a vital part of the development of an understanding of global warming. From the more abstract point of view of astronomy and our understanding of our place in the Universe, it means that when we have telescopes in space good enough to analyse light from planets orbiting other stars, we will be able to tell whether those planets are dead or alive, without ever visiting them. Gaia theory says that living planets anywhere in the Universe will be distinguished by atmospheres rich in gases like oxygen and ozone (see Chapter Nine).

But it would take a whole book to describe the nature of life on Earth, and its relationship with the physical environment. What we can do here is describe that physical environment from an astronomical point of view.

Calendrical Conundrums

It is generally best not to take things at face value in astronomy. To the ancients, it was obvious that the Moon, Sun and stars all go round the Earth, which was at the centre of the Universe. Today, even though 'everybody knows' that the cycle of day and night is not caused by the Sun going round the Earth, but by the Earth spinning on its axis, it turns out that you have to be careful to spell out what you mean by a 'day'.

We all know that there are 24 hours in a day, by definition. Or are there? The natural way to measure this is from noon on one day, when the Sun is at its highest in the sky, to noon on the next day. You might think that this corresponds to the time it takes for the Earth to turn once on its axis. But that is not quite the case. Because the Earth has moved on a little bit in its orbit between one noon and the next, it has to turn a little bit more before the Sun is at its highest in the sky over the same point. The interval between two noons is called the solar day. The time it takes for the Earth to turn once on its axis can be measured by looking at how the stars appear to move across the sky. If a particular star is at its highest in the sky at midnight, the time taken for it to appear in the same place again is called the sidereal day. The sidereal day is about 4 minutes shorter

Fig. 17. *Sidereal and solar days. The combination of the rotation of the Earth and its orbital motion means that there are two different ways to define a 'day'! See text for full discussion.*

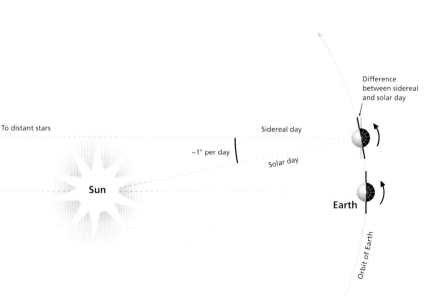

than the solar day. If the solar day is defined as 24 hours, the sidereal day, the time it takes for the Earth to turn once on its axis, is actually 23 hours 56 minutes and 4 seconds. So, strictly speaking, it is not even true to say that the Earth turns on its axis once a day (fig. 17).

The Earth does travel around the Sun once a year, by definition. It takes so long because the Sun is so far away. The average distance from the Earth to the Sun is 149,597,870 km – near enough 150 million km (93.2 million miles) – the astronomical unit of distance, or AU. Because the Earth's orbit is nearly circular, that means the distance it travels around that orbit once is roughly 940 million km (584 million miles) (remember that the circumference is π times the diameter), at an average speed of a little over 107,500 km (66,800 miles) per hour. But unfortunately there is not an exact number of days in the year. That is why we have leap years.

The length of a year is very nearly 365^1/$_4$ days, so to keep our calendar in step with the movement of the Earth around the Sun, every four years we have an extra day, 29 February, in years whose numbers divide exactly by 4. This would be perfect if the length of the year was exactly 365.25 days, but it is actually 365.242199 days, so by adding a whole extra day every four years we are slightly overcompensating. The way round this is that century years are only leap years if they *also* divide by 400. So the years 1700, 1800 and 1900 were not leap years, but the year 2000 was. In a final adjustment for the very long term, the years 4000, 8000 and 12000, which all divide by 4000, will *not* be leap years. This system is so accurate that the calendar will now only get out of step with the orbit of the Earth by a full day after 20,000 years, and if there is anybody still around then they can make a special one-off adjustment to their calendar.

Why should we care about keeping our calendar in step with the orbit of the Earth around the Sun? Because of the seasons. If we didn't have leap years, the calendar would gradually get out of step with the seasons, so that the first day of Northern Hemisphere spring (officially 21 March today) would occur later and later, until we had snow in the Northern Hemisphere in July and summer in January. It seems reasonable to stick with the relationship between months and seasons we are used to; but why are there seasons at all?

Fig. 18. Earth's South Polar icecap is shown by a mosaic of 11 photographs taken during a 10-minute period in December 1990.

Seasons and Super-seasons

We have seasons on Earth because the planet is tilted in its orbit. If you imagine a line from the Sun to the Earth, instead of this making a right-angle with a line through the North and South poles of the Earth it makes an angle of $66^1/2$ degrees. In other words, the Earth is tilted at $23^1/2$ degrees out of the vertical. Over the course of a year, the North Pole always 'points' in the same direction – as it happens, towards a particular star, known as the Pole Star. But as the Earth orbits round the Sun, this means that in one part of the orbit the Northern Hemisphere leans towards the Sun, while on the other side of the orbit the Northern Hemisphere leans away from the Sun – just as if the Sun had gone round behind the Earth. In between, there are two days in each year, on opposite sides of the orbit, when

Fig. 19. The Hubble Space Telescope seen after its second servicing mission in 1997. Hubble drifts 569 km (353 miles) above the Earth's surface, above the atmosphere, where it can see objects in space more clearly than telescopes on the ground.

the Sun is exactly on one side of the Earth, measured relative to a line through the poles. These are called the equinoxes. (See *fig. 20.*)

When the hemisphere is tilted towards the Sun, people living in that hemisphere see the Sun rise high in the sky, and the pole itself never sees the Sun set. It is summer. When the hemisphere is tilted away from the Sun, people living in that hemisphere see the Sun rising only a little way in the sky, even at noon, and the pole itself is dark 24 hours a day. It is winter. Because the Sun rises higher in the sky in summer, there are more hours of sunlight, and this is one reason why summer is warmer than winter. But when the Sun is high in the sky a patch of ground or sea gets more concentrated sunlight than the same patch does when the Sun is low in the sky. This is why the Sun feels hotter at noon than in the morning or in the evening. Imagine a 'beam' of sunlight 1 metre square arriving at the surface of the Earth at an angle. The shallower the angle the more the beam gets spread out over the surface. At high noon in summer, at the latitude of Spain each square metre of radiation from the Sun spreads out over $1^1/_4$ square metres of the Earth's surface; but at high noon in winter, the same square metre of solar radiation is spread over $2^1/_2$ square metres of Spain.

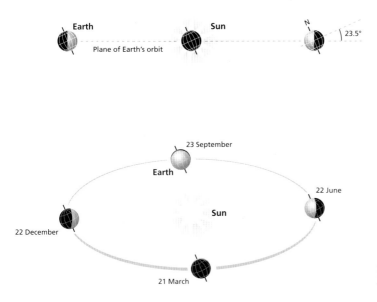

Fig. 20. *The orbit of the Earth around the Sun.*

The Sun is at its highest in the sky in the Northern Hemisphere on 22 June, the summer solstice; it is at its lowest noon altitude on 22 December, the winter solstice. This entire pattern of the seasons is reversed in the Southern Hemisphere. At the equinoxes, on 21 March and 23 September, everywhere on Earth the Sun rises exactly to the east and sets exactly to the west, and there are 12 hours of daylight and 12 hours of night. Because the Earth's orbit is not quite circular, we are at our closest to the Sun (perihelion) around 3 January, near the time of the Northern Hemisphere winter solstice; we are at our farthest from the Sun (aphelion) around 3 July. The Earth's orbit shifts slightly over thousands of years, so these dates are not absolutely fixed. The difference between aphelion and perihelion is not the cause of the seasons; in fact, it means that Northern Hemisphere winters are a tiny bit less cold than they would be if the Earth's orbit were circular, and southern hemisphere summers are a tiny bit cooler than they would be if that were the case.

All of this discussion of the seasons applies to the situation on Earth today. But the pattern of the seasons changes, on timescales of thousands of years, because of the gravitational influence of the Moon and other objects in the Solar System on the Earth. The shape of the Earth's orbit actually changes slightly, from more elliptical to more circular and back again. The Earth also wobbles, like a spinning top, as it orbits around the Sun, which means that the North Pole has not always pointed to the present-day Pole Star. This wobble, or precession, is very slow by human standards, and means that an imaginary line through the poles of the Earth traces out a circle on the sky once every 26,000 years. Four thousand years ago, in ancient Egypt when the pyramids were built, a star known as Thuban, in the constellation Draco, was the pole star. Most importantly, as far as the seasons are concerned, the angle of tilt of the Earth (its obliquity) changes. The present angle of $23\frac{1}{2}$ degrees (more exactly, 23.44 degrees) is about halfway between the extreme values that are reached, 22.1 to 24.5 degrees, and it takes about 41,000 years to complete each 'nod' up and down. Today, the obliquity is decreasing so the difference between the seasons is getting less.

The combined effect of all these changes is to alter the balance of heat between the seasons on Earth in a rhythmic way. We always receive the same amount of heat from the Sun averaged over the course of a year, but sometimes there are pronounced differences between summer and winter, and sometimes there is less difference between summer and winter. These rhythms are referred to as the Milankovitch Cycles, after the Serbian astronomer Milutin Milankovitch (1879–1958), who calculated how they affect the climate on Earth. The most important discovery is that when northern summers are cool this allows ice sheets to spread, because the snow that falls on the land in winter does not melt in summer. The Milankovitch Cycles – akin to super-seasons – explain why for the past million years or more the Earth has experienced a repeating pattern of ice ages lasting about 100,000 years separated by so-called interglacials, like the conditions on Earth today, which last for ten or fifteen thousand years.

Fig. 21. *The internal structure of the Earth.*

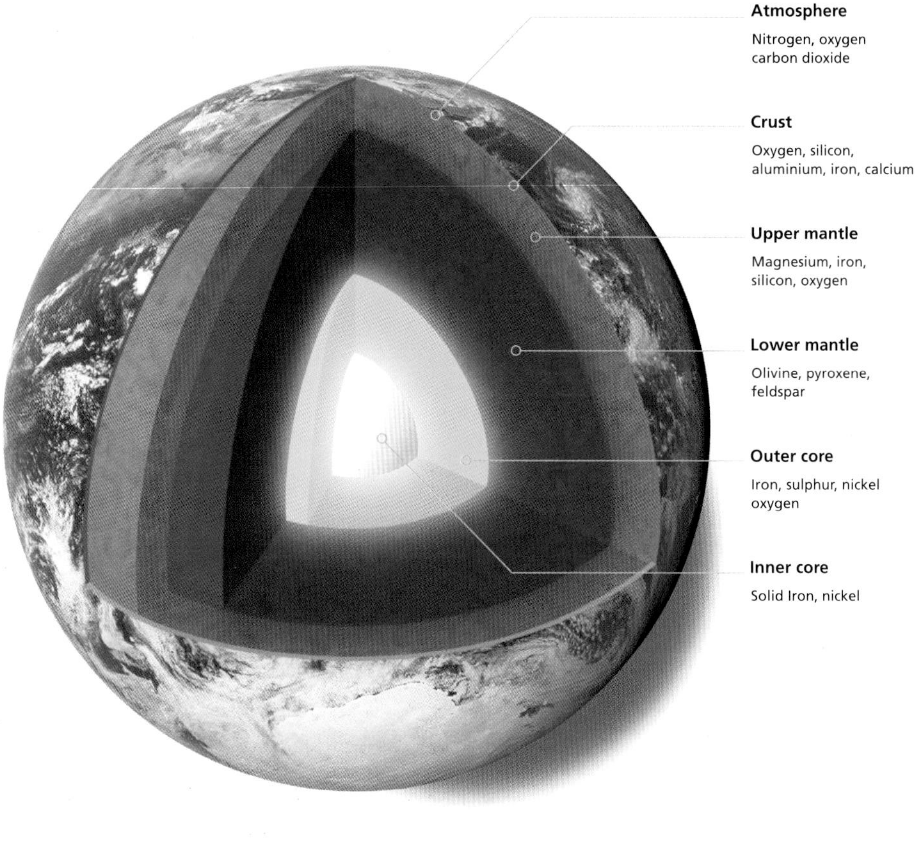

Atmosphere

Nitrogen, oxygen carbon dioxide

Crust

Oxygen, silicon, aluminium, iron, calcium

Upper mantle

Magnesium, iron, silicon, oxygen

Lower mantle

Olivine, pyroxene, feldspar

Outer core

Iron, sulphur, nickel oxygen

Inner core

Solid Iron, nickel

Inside the Earth

The 'solid' ground beneath our feet is almost as thin, compared with the diameter of the Earth, as the atmosphere. The cool crust of the Earth is actually a thin layer, like an eggshell, surrounding a bulk of hot material, which is divided into two main zones, reminiscent of the white and yolk of an egg (see *fig. 21*). The crust beneath the oceans varies in thickness from about 5 km (3.1 miles) to 11 km (6.8 miles), and has an average thickness of only 7 km (4.3 miles). Continental crust is thicker, with more variation from place to place. At its thinnest, it is no more than 20 km (12.4 miles) thick, but in the Himalayan region it is 90 km (55.9 miles) thick. The average is about 35 km (21.7 miles).

Like an eggshell, the thin crust can crack, especially under the oceans where it is thinnest. One major crack runs down the Atlantic Ocean, where there is a line of underwater volcanic and earthquake activity. Molten material from below the crust wells up through the crack and pushes out on either side as it sets to make new crust. This is pushing the ocean wider at a rate of a couple of centimetres a year.

But the Earth is not getting bigger, because as fast as new crust is being made in some places, in other places, particularly down the western side of the Pacific Ocean, thin ocean

Fig. 22. The principal tectonic plates. These make up the surface of the Earth, like a cracked eggshell.

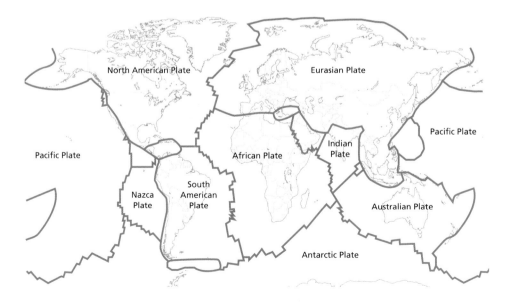

Fig. 23. *Volcanic activity shown in a lava channel in Hawaii, 1984.*

crust is being pushed under the thicker continental crust, forcing up mountain ranges and volcanic islands like those of Japan. This is called subduction. So although the Atlantic Ocean is getting wider, the Pacific Ocean is getting narrower.

The different pieces of cracked crust are called plates, see *fig. 22*. The plates jostle against one another and move around the surface of the Earth, carrying continents along with them, in a process known as continental drift. When continents collide, huge mountain ranges are pushed up-ward in the collision; the Himalayan mountains are so big because India is moving northward and ploughing into the main mass of Asia. The top of Mount Everest used to be the bottom of an ancient sea that has been squeezed out of existence.

This whole package of seafloor spreading, continental drift and subduction is known as plate tectonics. It seems to be unique to the Earth, at least in this form, in our Solar System. It all takes place within 0.6 per cent of the Earth's volume and 0.4 per cent of its mass. The total mass of the Earth is 5976 billion billion tonnes, and its volume is 1083 billion cubic kilometres, so its overall density is just over five and a half times the density of water.

Although we cannot see what lies beneath the crust, geophysicists are able to probe the internal structure of the Earth by analysing the way vibrations triggered by earthquakes are reflected by different layers beneath our feet. The earthquake waves go down deep into the Earth and are deflected by the different layers before bouncing back up to the surface and shaking the scientists' seismometers. The way the waves move shows that beneath the crust there is the mantle, analogous to the white of an egg, extending to a depth of nearly 2900 km (1800 miles). The mantle contains two-thirds of the Earth's mass, in just over four-fifths of the volume of the planet. Deeper still, the core of the Earth, analogous to the yolk of an egg, extends all the way from 2900 km (1800 miles) to the centre, at a depth of just under 6400 km (3980 miles). This represents the remaining fifth of the Earth's volume, and nearly a third of its mass. The outer part of the core, down to a little over 5000 km (3110 miles), is thought to be liquid, but the inner core seems to be solid. The mantle is also solid, but hot enough to flow very slowly, like warm glass, and have convection currents that stir it up over long periods of time. It is mostly made of silicates. But the core is mostly made of iron and nickel, heavy elements that settled in the core when the young Earth was molten.

In the core, the density is 13.5 times the density of water, falling to between 5.5 and 3.5 times the density of water in the mantle and about three times the density of water in the crust.

The Magnetic Earth

The Earth formed, along with the Sun and the other planets, about 4.5 billion years ago. Details of the way it formed are intimately linked with the formation of the Moon, and will be discussed in the next chapter. It has been cooling ever since, but not as fast as a cooling lump of iron the size of the Earth, because the core also contains radioactive elements, in particular uranium, thorium and a particular type of potassium. These release heat as they decay into more stable forms. Near the surface of the Earth, the temperature rises dramatically with depth, at a rate of 3 °C (5.4 °F) for every 100 metres (109 yards); but the rate of increase slows down so that at a depth of 150 km (93.2 miles) the temperature is about 1100 °C (2012 °F). By a depth of 650 km (403 miles), the temperature is up to 1900 °C (3452 °F), but the rest of the rise is fairly sedate, with the temperature in the inner core lying between 4000 °C (7232 °F) and 5000 °C (9032 °F). When all the radioactive elements have decayed, in something over 10 billion years from now, the Earth's interior will cool completely and tectonic activity will stop.

Because the core is rich in iron and nickel, it is a good conductor of electricity, and acts as a generator for the Earth's magnetic field. Nobody knows exactly how the magnetic field is generated, but it clearly has something to do with the presence of swirling currents of molten metal in the outer core, stirred up by the rotation of the Earth. The magnetic field of the Earth extends far out into space, forming a zone around the planet called the magnetosphere. On the side of the Earth nearest the Sun, the magnetosphere is squashed by a stream of particles from the Sun called the solar wind (see page 75 and *fig. 48*, page 78). As a result, it only extends out to about ten times the radius of the Earth. On the side away from the Sun, however, the main

Fig. 24. Aurora Borealis seen in Iceland, 13 March 2005. The Moon shines beneath the glowing arc of this Aurora display. The orange glow on the left side of this image is the lights of Reykjavik.

magnetosphere extends out as far as sixty Earth radii, with a tail stretching away hundreds of Earth radii into space. The electrically charged particles of the solar wind are deflected by the Earth's magnetic field and funnel down towards the north and south magnetic poles, where they interact high above the ground to produce the spectacular coloured displays of the aurorae – the northern and southern lights (fig. 24).

Without the protection of the magnetosphere, particles of the solar wind, and similar particles from outer space known as cosmic rays, would penetrate to the surface of the Earth, with damaging effects on life. This seems to have happened many times in the past. When molten magma flows from a volcano and sets to make rock, it seals in a record of the Earth's magnetic field at the time it was laid down. This record of fossil magnetism reveals that from time to time the magnetic field dies away to nothing, then builds up again, either in the same direction or with the magnetic poles reversed. Over the past 85 million years, the field has reversed nearly two hundred times. These magnetic reversals often correspond to times when there have been 'extinctions' of life on Earth, when an unusually large number of species, compared with the long-term average, disappear from the fossil record; but they are by no means lethal to all forms of life. At present, the strength of the Earth's magnetic field is decreasing at a rate which would see it disappear by about the year 4000; but it is quite possible that it will start building up its strength again before then.

The edge of the magnetosphere marks the final boundary between the Earth's sphere of influence and outer space; coincidentally, our Moon orbits at a distance of just over sixty-one Earth radii, near the edge of the magnetosphere proper. But although the Moon deserves a chapter of its own, not least as the first stepping stone from the Earth out into the Universe at large, its story is inextricably linked with the story of the Earth; although lifeless itself, it is responsible for many of the features of our planet that make the Earth a comfortable home for life.

Fig. 25. *Volcanic activity viewed from space. Mt Etna's thick, yellowish-brown plume billowing from the mountainside.*

Chapter_Two

Moon

Our Moon is one of the most remarkable objects in the Solar System. With a diameter of 3476 km (2160 miles), it is roughly a quarter of the size of Earth. That makes it much bigger in proportion to the size of the planet than any of the other moons of the planets orbiting the Sun. Because of this, some astronomers refer to the Earth–Moon system as a double planet – even though, because volume is proportional to the cube of the diameter, the volume of the Moon is just under 2 per cent of the volume of the Earth (fig. 27). The mass of the Moon is just 1.2 per cent of the mass of the Earth, about one-tenth of the mass of the planet Mars, but nearly a quarter of the mass of the planet Mercury. If it were not bound by gravity in an orbit around the Earth, the Moon would definitely count as a planet in its own right.

Those gravitational bonds keep the Moon orbiting at an average distance from the Earth of 384,410 km (238,860 miles). This is the distance from the centre of the Earth to the centre of the Moon; the average distance from the surface of the Earth at the equator to the surface of the lunar equator is 376,284 km (233,810 miles). However, both the Earth and the Moon are in fact orbiting around the balance point in the Earth–Moon system, its centre of mass. This is like the balance point of a see-saw, and just as the weight of a child on one end of the see-saw can be balanced by an adult sitting on the other side but much closer to the balance point, so the balance point of the Earth–Moon system lies much closer to the centre of the Earth than it does to the centre of the Moon. It actually lies below the surface of the Earth, about three-quarters of the way (4700 km, 2920 miles) out from the centre.

Changing the analogy, like a light female ice skater being swung around by her larger male partner, the Moon follows

Fig. 26 (overleaf) The large crater Tsiolkovsky (with a diameter of 240 km (149 miles)), as photographed by the astronauts during the Apollo 8 lunar orbit mission, looking east toward the lunar horizon.

a wide orbit around the point of balance, but, like the male skater holding on to her, the Earth is also being swung around in a smaller circle. It takes 27.3 days to complete each orbit, measured against the background stars; this is called the sidereal month. However, because of the way the Earth–Moon system moves around the Sun, the time taken for the Moon to run through its complete cycle of phases, from New Moon through Full Moon and back to New Moon again, is 29.5 days: the synodic month.

Fig. 27. The Earth–Moon 'double planet', photographed by the Galileo spaceprobe in a fly-by in 1992.

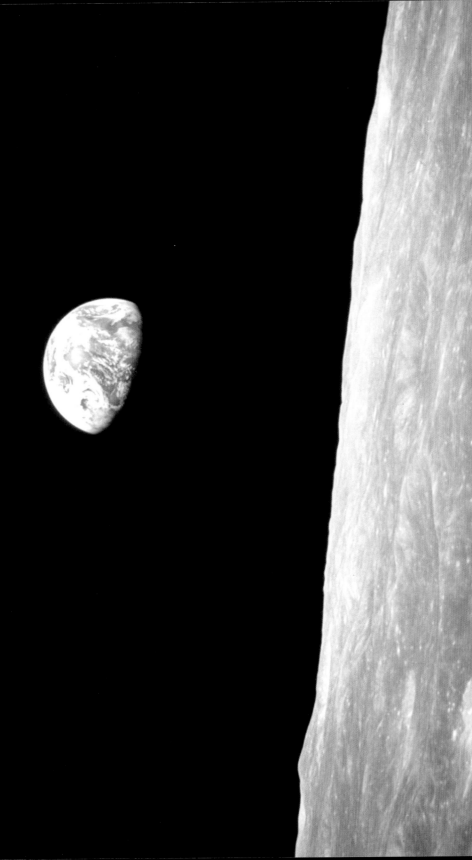

Old and New Moon

The phases are simply caused by the changing orientation of the Moon relative to the Sun and the Earth as it moves round its orbit. At Full Moon, the Sun is on the other side of the Earth from the Moon, and shines directly on to the lunar surface visible from Earth. At New Moon, the Sun is on the other side of the Moon from the Earth, and shines on the side we cannot see; we only see a sliver of illumination around one edge. In between phases correspond to in between orientations. Contrary to popular myth, it is not true that there is a permanent 'dark side' to the Moon. The Moon does keep the same face directed towards the Earth, because tidal effects have locked its rotation in step with its orbit (the Moon turns once on its axis every time it orbits once around the Earth – once every 27.3 days) and we always see the same face of the Moon, but the other side is only completely dark when the face we see is fully lit (at full Moon). From the surface of the Moon, you would see a cycle of day and night as the Sun rose and set, just as it does on Earth, but more slowly. A 'day' on the Moon lasts for just over 29 days and 12 hours, divided equally between light and dark. During the long night the temperature falls to −170 °C (−274 °F), but at lunar noon the temperature reaches 100 °C (212 °C), hot enough to boil water.

The oldest lunar rocks, dated by radioactive techniques, are some 4 billion years old; older than the oldest rocks found on Earth. This is because the Moon is not active in the way the Earth is, there are no tectonic processes reworking old rock into new forms. There was some activity when it was younger and still hot inside; the lava flows that were produced can still be seen in the lowlands of the Moon. It is now cold inside and almost inert, although there may still be occasional traces of volcanic activity. The interior structure of the Moon has been probed by studying seismic activity – moonquakes – using instruments left behind by the Apollo astronauts in the 1970s. Beneath a crust 60 to 100 km (37 to 62 miles) thick there is a layer of slightly denser rock about 1000 km (620 miles) thick, surrounding an inner core less than 700 km (434 miles) in radius that may be as hot as 1500 °C (2732 °F), warm enough to melt rock, but lacks the iron and other heavy elements found in the Earth's core.

Fig. 28 'Earthrise'. This view greeted the Apollo 8 astronauts as they came from behind the Moon after the lunar orbit insertion burn.

The oldest rocks on the Moon are found in the lunar highlands, where they have not been covered by lava. The highest mountains on the Moon are in the Leibnitz range just visible from Earth, near the South Pole, where they reach more than 8 km (4.9 miles) above the surrounding lowlands – there is no 'sea level' on the Moon to measure heights from since there are no seas and no water.

There is also no atmosphere on the Moon, which is why the craters that cover its surface have remained so distinct – they have not been eroded away by the action of wind

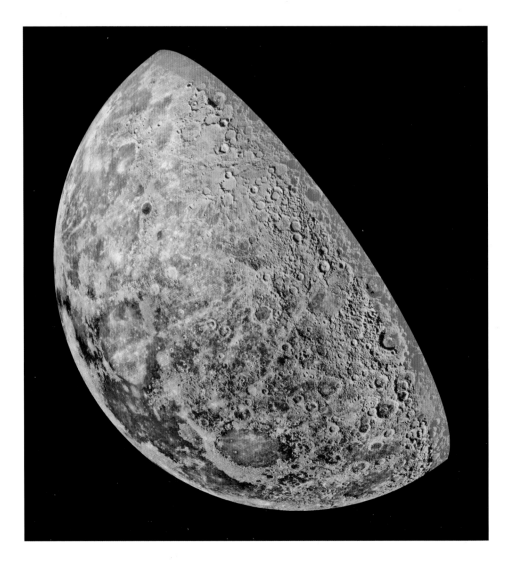

and weather. The largest craters are more than 200 km (124 miles) in diameter and the smallest ones visible from Earth are about a kilometre across; but they actually extend all the way down to pits about a centimetre across. They are all a result of impacts by meteorites. It is estimated that the visible side of the Moon alone is covered with more than 300,000 craters, each at least a kilometre across.

The lack of atmosphere is because the Moon is so small – its gravitational pull is not strong enough to hold on to an atmosphere. The force of gravity at the surface of the Moon is only 16.6 per cent of the equivalent force at the surface of the Earth, so on the Moon you would weigh only one-sixth as much as you do on Earth. Overall, the density of the Moon is 3.3 times the density of water, similar to the density of the Earth's crust. The chemical composition of the Moon is also similar to that of the Earth's crust, and these are important clues to the origin of the Earth's unusually large natural satellite.

Fig. 29. False-colour mosaic taken by Galileo's space-probe imaging system in 1992. The part of the Moon visible from Earth is on the left. The mosaic shows compositional variations in the Moon's northern hemisphere. Bright pinkish areas are highlands, such as those surrounding the oval lava-filled Crisium impact basin toward the bottom of the picture. Blue to orange shades indicate volcanic lava flows. To the left of Crisium, the dark blue Mare Tranquillitatis is richer in titanium than the green and orange maria above it. Thin mineral-rich soils associated with relatively recent impacts are represented by light blue colours; the youngest craters have prominent blue rays extending from them.

The Origin of the Moon

Nobody can be sure how the Moon formed, because it happened so long ago; but by analysing the chemical composition of Moon rocks, studying the way the Moon and planets have been cratered, and using computer simulations, astronomers have come up with a 'best buy' scenario. It is called the 'Big Splash' (*fig. 30*).

We do know that the Sun and the planets of the Solar System, including the Earth, formed from a collapsing cloud of gas and dust a little more than 4.5 billion years ago (see Chapter Seven). By about 4.6 billion years ago, the Sun was shining at the centre of the Solar System, and the Earth was one of several balls of rock orbiting around the young Sun. These primordial proto-planets had grown by sweeping up smaller objects; the impact of this rain of cosmic debris generated enough heat for the whole planet to melt, with iron and other heavy elements sinking into the core. The surface of the planet cooled and solidified as the bombardment of meteorites eased off, and began to form a thick crust around the molten interior. But just as the Earth was beginning to settle down in this way, it was struck a glancing blow by one of the other rocky objects that had been orbiting the Sun with it. This was a proto-planet about the size of the planet Mars, with roughly one-tenth of the mass of the Earth (ten times the mass of the Moon).

This impact would have melted both the incoming object and the original crust of the Earth. Heavy metals like iron from the core of the proto-planet would have sunk down into the core of the Earth, but lighter material, including a great deal of the Earth's original crust, would have been flung out into space in the Big Splash. Most of it escaped entirely from the Earth's gravitational pull, but some settled into a ring around the Earth. That ring of material cooled and coagulated to form the Moon, which is why the composition of the Moon is similar to that of the Earth's crust. The Earth was left with a very thin crust, much thin-

Fig. 30. The 'Big Splash'.

ner than it would otherwise have had, allowing the kind of

Fig. 31. *An oblique view of Copernicus and the Carpathian Mountains, obtained in 1966 by the spaceprobe Lunar Orbiter II.*

tectonic activity that we described in the previous chapter to shape the surface of the planet we live on.

Something similar, but if anything even more dramatic, may have happened to the planet Mercury (see Chapter Four). The Moon is like the crust of the Earth without a core, but Mercury is like the core of the Earth without a crust. A possible explanation is that early in its history Mercury was also struck by another proto-planet, but in a head-on collision, not a glancing blow. In a head-on collision, the iron-rich cores of the two objects would have been driven into one another, but the light material of their crusts would have melted and been blown away into space.

Since the Moon formed, it has continued to be bombarded by impacts from space, producing its present, heavily cratered surface. This rain of meteorites has eased off since the Solar System formed, but in the early part of the Moon's history there were several extremely large impacts by bodies tens of kilometres across that produced basins up to 25 km (15.5 miles) deep that have been filled in by molten lava to produce dark regions known as 'seas', or maria, although they do not contain water – the Moon is a dry, airless desert. Some of these maria are hundreds of kilometres

5__Timothy Ferris, *The Whole Shebang: A State of the Universe(s) Report* (London: Weidenfeld & Nicolson, 1997).

across, and pockmarked by smaller craters produced by lesser impacts over the eons.

Because the Earth was struck a glancing blow, it was sent spinning much more rapidly than it does today. The young Moon was also much closer to the Earth than it is today. Just after it formed, the distance from the Moon to the Earth's surface was only about 25,000 km (15,530 miles). Even a billion years later, 3.5 billion years ago, when there were already single-celled forms of life on Earth, the Moon was only about one-sixth as far from the Earth as it is today, about 66,000 km (41,000 miles), so that it would have looked six times bigger on the sky. It is still retreating, at a rate of 4 cm (1.5 inches) every year. There is no way to tell exactly how fast the Earth was spinning just after the Big Splash, but it was probably once every couple of hours.[5] By 400 million years ago, the fossil record shows that it was spinning once every 22 hours, so there were 400 days in the year; it is still slowing down now, at a rate of about 1.5 milliseconds per century. The Earth's spin has slowed down, and the Moon has gradually retreated from the planet, because of the effect of tides.

Time and Tide

Tides are caused by the gravitational influence of the Moon and the Sun – and, in principle, gravitational forces from

Fig. 32. Apollo 17 astronauts took this image of the Moon's Taurus-Littrow valley, with the lunar roving vehicle near the rim of Shorty crater. In the distance are the mountain-like massifs that define the valley.

other planets, but these are too small to notice. They are a result of the difference, in percentage terms, of the strength of the Moon's or Sun's gravitational pull from one side of the Earth to the other. This is called the gravitational gradient. The actual force exerted on the Earth by the Sun is 179 times the gravitational force exerted by the Moon, but because the Sun is 389 times farther away than the Moon, the gradient it produces is much less, and as a result the tides raised by the Sun on Earth are only 46 per cent as big as the ones raised by the Moon. We will only describe the lunar tides, because the solar tides occur in exactly the same sort of way (fig. 33). Sometimes, at New Moon and Full Moon, the lunar and solar tides add together, and we have very high spring tides (so-called because they spring up very high, not because they occur in the season called spring). Sometimes, halfway between the spring tides, the two effects oppose each other and we have lower, or neap, tides.

A proper explanation of tides involves a mathematical discussion of gradients and forces, but we can provide a good picture of what is going on by thinking in terms of centrifugal force, a concept familiar in everyday life.[6] This is the force you feel flinging you outwards when a car goes round a tight bend at high speed. Physicists sometimes call it a 'fictitious' force, because it is an effect caused by motion, not a force like gravity that is there all the time; but what matters is that, from the point of view of the

6__For further reading on gradients and forces see Open University (Oceanography Course Team), *Waves, Tides and Shallow-Water Processes* (Milton Keynes: Open University, 2000).

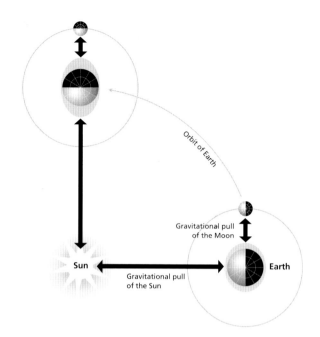

Fig. 33. *The gravitational pull. Both the Sun and the Moon raise tides on the Earth. Sometimes they work together, making 'spring' tides; sometimes they oppose one another, making 'neap' tides.*

rotating object, things are flung outward from the centre of rotation. In the case of the Earth and the Moon, the centre of rotation is the centre of mass that, as we have seen, is 4700 km (2920 miles) out from the centre of the Earth and exactly under the Moon.

On the side of the Earth directly under the Moon, the Moon's gravitational influence is strongest, and pulls the water of the oceans (and, in fact, the land itself, to a lesser extent) up in a hump towards it. On the far side of the Earth from the Moon, the Moon's gravitational influence is weakest, but it acts to pull the water down towards the Earth. But, at the same time, the Earth is rotating around the centre of mass of the Earth–Moon system. This produces a centrifugal effect that acts to fling water outwards. On the side nearest the Moon, this produces a small effect, because on that side the surface of the ocean is closest to the centre of rotation. This small effect makes the tide a tiny bit higher. But on the other side, away from the Moon and at the greatest distance from the centre of rotation, the centrifugal effect is biggest, and makes a high tide in spite of the weak tendency of the Moon's gravity to flatten the hump. The overall result is that there are two bulges of water, on opposite sides of the Earth. As the Earth rotates once every

24 hours, each part of the planet experiences both bulges once every day, giving us two high tides and two low tides.

In fact, the tidal bulges do not lie directly on a line joining the centre of the Earth to the centre of the Moon. Because of the rotation of the Earth, the bulge 'under' the Moon is actually carried slightly ahead of the point directly under the Moon. In the open ocean, high tide occurs about 12 minutes before the Moon is at its highest in the sky. It is the gravitational force between this leading bulge and the Moon that gradually slows the rotation of the Earth and moves the Moon outward in its orbit. In effect, energy is transferred from the Earth to the Moon.

Twice each month, at the times of New Moon and Full Moon, the Sun's effect adds to the lunar effect, and the tides 'spring' up to their greatest height – they are known as spring tides. At the two half-moon phases of the lunar cycle, the lunar and solar influences work in opposition to one another and there is the smallest difference between high tide and low tide – these are called neap tides. Because the Moon orbits the Earth once every 27.3 days, the tides do not repeat with an exact 24 hour rhythm. By the time the Earth has turned once on its axis, the Moon has moved on a little in its orbit. The Moon is in the same position relative to a point on the surface of the Earth every 24 hours and 50 minutes, so the tides occur 50 minutes later each day.

All of this applies to tides in a perfectly uniform ocean covering the entire surface of the planet. On the real Earth, land masses make barriers that disturb the flow of tides and alter the pattern locally. The Mediterranean Sea only has very small tides, because it is difficult for water to flow in and out of the narrow Strait of Gibraltar. On the other hand, in the Bay of Fundy, in Nova Scotia, Canada, the shape of the coastline causes a difference between high and low tide of as much as 15 metres (fig. 34). In some places such as Karumba, in Northern Australia, there only seems to be one tide a day, because of the way all the effects interact.

Tides have been a major influence on the Earth – slowing down its rotation, causing erosion, and providing a region of shoreline covered by water twice a day and left dry twice a day, in which living things evolved the capacty to breathe air and move on to the land. The Moon has also had another role in making our planet a suitable home

for life, by acting as a kind of stabiliser that stops the Earth from toppling over in space.

Earth's Stabiliser

Because of centrifugal effects the spinning Earth is not precisely spherical, but has a slight bulge at the equator. It is this bulge that provides the 'handle' for the gravity of the Sun and other bodies in the Solar System to tug on the Earth and cause the precession described in Chapter One. But this precession isn't the only result of the tugging at the Earth's equatorial bulge. Any planet in such a situation is tugged on in different ways by many forces, and some of these forces can get in step with the overall motion of the planet to produce a much more dramatic change. This process is called a resonance. It is like the way in which a series of small pushes on a child's swing can make the swing go higher and higher, if each little push is timed to produce the maximum effect. When the pushing doesn't happen at precisely the right time, the swing simply jiggles about over a small distance.

The best example of how such resonances work in the Solar System is provided by the planet Mars, which, as it happens, has roughly the same axial tilt as the Earth today, close to 25 degrees. The Earth's tilt is thought to be a result of the impact that led to the formation of the Moon, which has been frozen in, as we shall explain, by the presence of the Moon. But the tilt of Mars today is not the same as it has been since the planets formed. Mars precesses with a period of 157,000 years, compared with the Earth's precession period of 26,000 years, but it also experiences a resonance that causes the tilt to change between 15 degrees and 35 degrees over a period of several hundred thousand years. For most of that time, the forces are out of step and there is no resonance; but for a short time they are in step and the planet topples over before settling into a new orientation for a few hundred thousand years. It is quite easy to calculate this – if you use a computer to do all the tedious arithmetic.

Even more dramatically, more detailed calculations carried out in 1993 by the astronomer Jacques Laskar revealed that over periods of tens of millions of years the tilt of Mars varies between zero (upright) and 60 degrees, so that at

Fig. 34. Bay of Fundy tidal range. These images from the Advanced Spaceborne Thermal Emission and Reflection Radiometer (ASTER) on NASA's Terra satellite show the dramatic difference in the amount of water-covered land at the head of the southeast corner of the bay during a high tide on 20 April 2001, and a low tide on 30 September 2002. Vegetation is green, and water ranges from dark blue (deeper water) to light purple (shallow water).

times it is almost lying on its side in its orbit. These changes do not have any regular cycle, but occur unpredictably, in a chaotic fashion. One effect of such an extreme tilt would be to warm the polar ice caps on Mars, perhaps releasing liquid water to flow over its surface.

If such a thing happened to the Earth and the tilt changed to 90 degrees, not just the poles but each hemisphere would experience six months of 'night' followed by six months of 'day'. But it would get so hot near the poles during the long day – between 50 °C (122 °F) and 100 °C (212 °F) – that they would never cool off completely during the night, so the polar regions would never freeze. Instead, the region around the equator would be covered in ice all the year round. It is quite possible that life could exist and evolve on such a planet. But the real problem occurs when the planet topples from one orientation to another, which can happen in the span of a few tens of thousands of years – tens of thousands of years of climate change and turmoil. Life would survive, but civilisation might not. This doesn't happen to the Earth because of our large Moon; so maybe the only reason we are here to notice such things is because of the existence of the Moon.

Laskar calculated that without the presence of the Moon the Earth would wobble even more violently than Mars does, toppling up and down between zero degrees and 85 degrees with a rhythm only a few tens of millions of years long. The reason the Moon saves us from this fate is that it makes the Earth precess much more rapidly than Mars, once every 26,000 years, and this is too fast for any of the other forces we have mentioned to get in step with the precession to cause a resonance and make the Earth tip over. All that happens is that the tilt shifts over a range of about 2.5 degrees with a rhythm roughly 41,000 years long – and as we have seen, even that small change is enough to play a big part in the Milankovitch Cycles of ice ages.

The Moon has probably stabilised the Earth in this way for billions of years; but there is nothing special about the particular angle by which the Earth is tilted out of the vertical, and if the tilt had been bigger at the time the Moon formed, then the Earth would have been stuck in that orientation, with a very different pattern of climate. But only a really large Moon can do the trick. Even if the Moon had been half as big as it actually is, it would not have been large enough to do the job of stabilising the Earth and stopping its climate from experiencing sudden and extreme changes. Although Mars does have two moons, they are tiny, and have no influence on its wobbles.

The Moon cannot, though, stabilise the Earth forever. As it recedes slowly from us, at a rate of 4 cm (1.5 inches) every year, its gravitational influence on the Earth is gradually getting weaker. By about 2 billion years from now – less than half of the time that has already passed since the Earth formed – it will be too weak to stop the tilt of the Earth changing, and there will be nothing to prevent our planet toppling in the way Laskar has described.

Eclipses

The fact that the Moon is receding from the Earth means that, as seen from the surface of our planet, it covers a smaller part of the sky today than it did long ago, and in the future it will look smaller still. This makes for one of the oddest

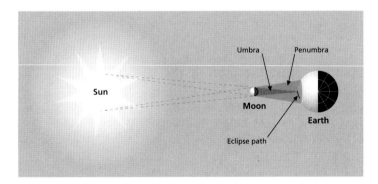

Fig. 35. *A solar eclipse. This occurs when the Moon's shadow passes across the surface of the Earth.*

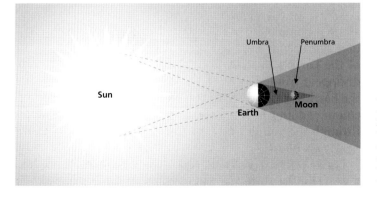

Fig. 36. *A lunar eclipse. Partial lunar eclipses and penumbral eclipses occur when the Moon is partly shadowed. A total lunar eclipse occurs more rarely, when the Moon is entirely within the Earth's shadow.*

coincidences in astronomy – just at the time when there are astronomers around to notice such things, the Sun and the Moon both look the same size on the sky as viewed from Earth. The Sun is nearly 400 times bigger in diameter than the Moon, but it is also nearly 400 times farther away from us. And because the Moon and Sun both appear the same size from Earth, we have spectacular solar eclipses.

We also have lunar eclipses (fig. 36), but these are less remarkable. They occur when the Moon passes through the shadow of the Earth, and the shadow of the Earth is rather big – it stretches out in a cone of complete darkness, called the umbral cone, for nearly 1.4 million km (870,000 miles) behind the Earth to a point in space. Around this zone of darkness there is an even bigger zone of partial darkness, called the penumbral cone. In the penumbral cone, light from the Sun is only partially blocked; part of the Sun's disc would be visible, with the Earth blocking out the rest.

At the distance of the Moon from the Earth, the penumbral cone is about 16,400 km (10,190 miles) wide and the umbral cone is about 9200 km (5720 miles) wide: more than two and a half times the width of the Moon. You might think that as a result the Moon would pass into the shadow, and be eclipsed, every time there is a Full Moon, with the Sun, Earth and Moon in a line. This doesn't happen every month because the orbit of the Moon is actually tilted slightly relative to the Earth's orbit around the Sun, at an angle of about 5 degrees. Most of the time, the Moon passes just above or just below the Earth's shadow and is not eclipsed. A lunar eclipse only happens if a Full Moon occurs just at the time the Moon is crossing the Earth's orbit, at a point known as a node. And, of course, it happens in the middle of the night.

When all the elements are in place for an eclipse, the Moon is completely in the umbral region for up to 104 minutes, and spends another couple of hours getting into and out of the Earth's shadow. But the Moon does not completely disappear from view, because sunlight is refracted around the Earth's atmosphere to give the Moon a dull red, coppery colour. Adding in the time it takes for the Moon to pass through the penumbral region, a lunar eclipse can last for nearly six hours.

Lunar eclipses are quite rare – there are only about two a year. But because the Moon is visible from everywhere

on the right side of the Earth, it isn't difficult to see one, if the sky is clear and you don't mind missing some sleep (fig. 37). Solar eclipses are actually more common than lunar eclipses, and there can be as many as five in a single year. But because any particular solar eclipse is only visible from a small part of the Earth's surface, different for every eclipse, you are unlikely to see one in any particular year unless you are willing to travel to the right part of the globe. But at least they happen in daytime.

Just as a lunar eclipse occurs when the Earth's shadow falls on the Moon, so a solar eclipse occurs when the Moon's shadow falls on the Earth (fig. 35). Once again, this can only happen at a node, but this time around midday, at a time of

Fig. 37. The partially eclipsed Moon, viewed over the desert landscape of Western Australia.

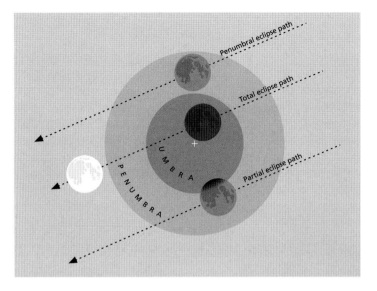

Fig. 38. Lunar eclipse paths.

New Moon when the Moon lies exactly on a line between the Sun and the Earth. Because the Moon is so much smaller than the Earth, its shadow is much smaller, and the umbral cone barely reaches the surface of our planet even at the best alignments. The Moon's orbit is not precisely circular, so sometimes it is closer to the Earth's surface than at other times. The Earth–Moon distance varies by about 10 per cent. If a solar eclipse occurs when the Moon is at its closest to us, the maximum width of the umbra at the surface of the Earth is just 270 km (168 miles), but the penumbra (the region in which you can see a partial eclipse) is much wider, close to 7000 km (4350 miles). Both the umbra and the penumbra sweep out a curving path several thousand kilometres long across the surface of the Earth, at a speed greater than 1500 km (930 miles) per hour, and everywhere along that strip the zone of totality is never more than about 270 km (168 miles) wide.

When the Moon is at its farthest from us, it looks a little smaller and cannot completely cover the Sun's disc. At such times, the tip of the umbral cone is actually a little more than 20,000 km (12,430 miles) above the Earth's surface; the result is that observers on the ground see a ring of light around the Moon; this is called an annular eclipse. But when the Moon is a little closer to us it looks almost the same size as the Sun from the surface of the Earth; the result is a less common total solar eclipse in which the Moon exactly covers the face of the Sun for a brief time – typically, a few minutes. Annular eclipses outnumber total eclipses in the ratio 5:4. In the future, total eclipses will become more rare as the Moon recedes from the Earth; within about 1.25 million years, they will no longer happen at all.

Just before the Moon completely covers the face of the Sun, if conditions are right, observers on Earth see the last flash of sunlight shining round the edge of the Moon in a spectacular 'diamond ring' effect; a little later, they may see the Moon surrounded by a necklace of lights, called 'Bailey's Beads', as the sunlight shines through valleys around the rim of the Moon. But as well as being a beautiful spectacle solar eclipses have practical value in astronomy. With the blinding light from the main disc of the Sun obscured in a total eclipse, it is possible to see a cloud of glowing gas, called the corona, around the Sun, as well as huge jets of material, known as prominences, leaping up from the Sun's surface to distances of thousands of kilometres. Observations of such phenomena made during eclipses have helped astronomers to develop their understanding of the Sun, and to appreciate that, although it is the dominating feature of our Solar System, it is in fact an ordinary star, nothing special in the Universe at large.

Fig. 39. *Lunar far side imaged by the Galileo spaceprobe in 1990.*

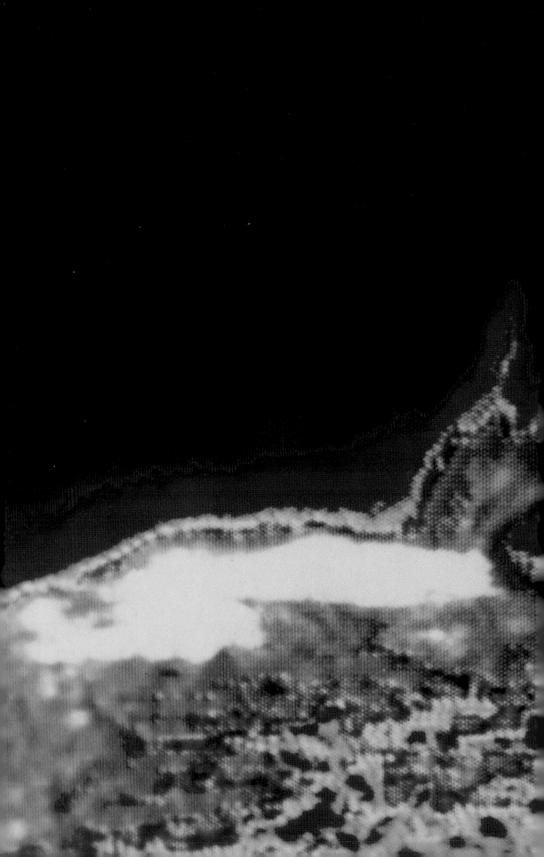

Chapter_Three

Sun

The Sun is by far the biggest thing in the Solar System, even though it is nothing special compared with other stars. The mass of the Sun is 330,000 times the mass of the Earth – that's near enough a third of a million times the mass of our planet. Everything else in the Solar System adds up to give a total mass of about 440 times the mass of the Earth. So very nearly 99.9 per cent of the mass of the entire Solar System is locked up in the Sun, which contains 745 times as much mass as all the planets put together. That leaves just over 0.1 per cent for everything else. Planets are like tiny specks of dirt in the emptiness of space around the Sun. The Latin name for the Sun, *sol*, is the root of the term Solar System.

The Sun looks so bright from Earth compared with other stars because it is so close to us. The next nearest star is some 270,000 times farther away than the Sun. Because the Sun is so bright as seen from Earth, like all authors of astronomy books we have to warn you never to look directly at the Sun, even during an eclipse; this can seriously damage your eyes and even cause permanent blindness.

The light from the Sun is made up of all the colours of the spectrum, so it would look white from space. It looks yellow from the surface of the Earth because the atmosphere of our planet scatters out some of the blue light, subtracting it from the Sun's light; this effect is particularly pronounced at sunset and sunrise, when the Sun appears orange or red. It is the scattered blue from sunlight that makes the sky look blue. It is no coincidence that the Sun radiates most of its energy in the region of the spectrum we call visible light; over millions of years our eyes have evolved to make best use of the light available.

The diameter of the Sun is 1,392,530 km (865,280 miles), about 108 times the diameter of the Earth (*fig. 41*), so its

Fig. 40 (overleaf)
A greatly enlarged false-colour image of a solar flare.

radius is also 108 times the radius of the Earth. Because volume goes as the cube of the radius,[7] and 100^3 is a million, the volume of the Sun is a bit more than a million times the volume of the Earth – to be precise, 1,295,000 times the volume of the Earth. Because the mass of the Sun is only a third of a million times the mass of the Earth, this means that in round numbers the density of the Sun is just one-third of the density of the Earth, only 1.4 times the density of water. But most of this mass is concentrated in a small central region of the Sun, called its core.

7_Specifically, volume $= {}^4/_3 \pi r^3$.

Inside the Sun

We know about the internal structure of the Sun in the same way that we know about the internal structure of the Earth and the Moon – by studying vibrational waves that travel through the interior and shake the surface. Just as earthquakes and moonquakes make the surfaces of these solid objects vibrate, so sound waves inside the fluid Sun make ripples on its surface. These ripples interact with one another to make waves in which patches of the Sun's surface move in and out, and these oscillations can be studied using spectroscopy and the Doppler effect (see Glossary). The patterns made by the oscillations reveal the internal structure of the Sun. This technique is known as helioseismology.

Fig. 41. The relative sizes of the Sun and Earth.

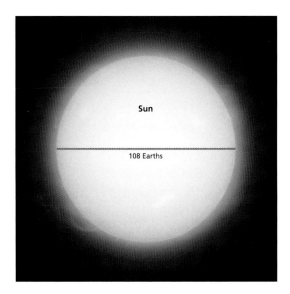

The Sun is composed of about 71 per cent hydrogen, just over 27 per cent helium and just under 2 per cent heavier elements. In a sense, it is a ball of hot gas; but this description hardly does justice to its internal structure.

From helioseismology and other studies astronomers know that the core of the Sun extends only a quarter of the way out from the centre to the surface, which represents just 1.5 per cent ($0.25 \times 0.25 \times 0.25$) its volume. Half of the Sun's mass is squeezed into this dense core. There, electrons are entirely stripped away from the outer parts of their atoms and the atomic nuclei are packed together so closely that the density is 160 times the density of water, or 12 times the density of lead. The pressure in the core is 300 billion times the pressure of the atmosphere at the surface of the Earth, and the temperature falls from some 15 million degrees Celsius at the centre of the Sun to 13 million degrees at the outer edge of the core.

Under these extreme conditions, at the very heart of the Sun a series of nuclear reactions takes place that results in the conversion of hydrogen nuclei into helium nuclei. This involves several steps that convert four hydrogen nuclei

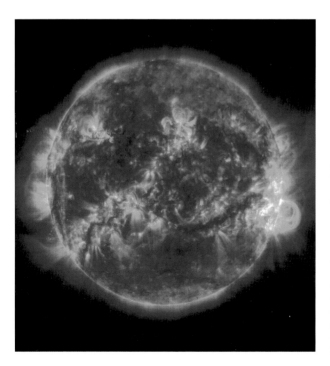

Fig. 42. The Sun as it would look to ultraviolet eyes. This 'false colour' image was recorded in 2002. The temperature of this material is about 1 million K in the lower corona. The large loop on the right arches over an active region. The extreme ultraviolet image enables us to see tight, loop-like magnetic fields that extend above the Sun's surface, around which charged particles are spinning.

(protons) into one helium nucleus (also known as an alpha particle). Each time this happens, 0.7 per cent of the mass of the original four atoms is released as energy, in line with Einstein's equation $E = mc^2$.

The process is quite rare, because protons each have a positive electric charge and tend to repel one another. It is only in occasional head-on collisions that they get squeezed close enough to one another to interact and fuse. In order to get an idea of how rare such collisions are we have to use some very large numbers, and it helps to use the standard mathematical notation in which 100 is represented by 10^2, 1000 by 10^3, and so on. In this notation, a number like 345,000 can be written as 3.45×10^5. When multiplying such numbers, you add up the powers (also known as indices) so that $10^5 \times 10^3$ is 10^8 and so on. For division, you subtract indices. It is important to remember that, for example, 10^4 is not twice 10^2 but a hundred times 10^2, because $10^4 = 10^2 \times 10^2$ and $10^2 = 100$.

It is estimated that there are about 10^{56} protons in the core of the Sun – that is, a 1 with 56 noughts after it. Out of these, only about 3.4×10^{38} fuse into helium nuclei each second. 3.4×10^{38} is a huge number to us, but it is only a tiny fraction of 10^{56}. Because $56 - 38 = 18$, it would take about 10^{18} seconds, which is about 100 billion (10^{11}) years, to use up every single proton in the core of the Sun. That will never actually happen, because long before then helium builds up like ash in the core and the Sun will undergo upheavals that we describe in Chapter Six. But we have spelled all this out not just because it gives an insight into how rare fusion is inside the Sun, but because the way of representing big numbers using indices will be useful when we look at the Universe beyond our Solar System.

There are so many protons inside the Sun that even though the process is quite rare, overall 5 million tonnes of mass is converted into energy and lost into space as radiation every second. In round numbers, every second 700 million tonnes of hydrogen in the core of the Sun is converted into 695 million tonnes of helium. But the Sun is so big that although matter inside it has been converted into energy in this way for 4.5 billion years, it has only lost 4 per cent of its original store of hydrogen nuclei so far. It has enough left to carry on burning it at the same rate for about another 5 or 6 billion years before the build-up of nuclear ash in the core causes problems. A star that is enjoying this period of quiet life, holding itself up against the inward pull of gravity by converting hydrogen into helium in its core, is said to be a main sequence star, or to be 'on the main sequence'.

Almost all of the energy released by nuclear fusion at the heart of the Sun is in the form of gamma rays, intense electromagnetic radiation even more powerful than x-rays. But the nuclear reactions also produce a flood of the lightest particles known (apart from photons, which have no mass at all), called neutrinos. Each neutrino has only about a millionth of the mass of an electron, and they do not carry much energy compared with the electromagnetic radiation released by the fusion process, but they have a curious property that makes them a unique probe of the solar interior.

Escaping from the Sun

Neutrinos are extremely reluctant to interact with other forms of matter, and if a beam of neutrinos like those being produced inside the Sun were to pass through a wall of solid lead three thousand light years long only half of them would be stopped by nuclei of lead atoms along the way. So the Sun is almost transparent to neutrinos, and neutrinos from the heart of the Sun travel out through it and across space at almost the speed of light without being impeded. Yet astronomers on Earth have been able to devise detectors sensitive enough to capture a few of the solar neutrinos passing through the Earth on their journey into space. The detection of solar neutrinos was one of the triumphs of late twentieth century astronomy, and combining these observations with the modern understanding of particle physics has enabled astronomers to open a new window, giving them a direct view of what is happening at the heart of the Sun today. This is particularly useful because studying light from the surface of the Sun can only tell us what was going on inside the Sun a long time ago.

The electromagnetic radiation produced by nuclear reactions in the inner core of the Sun starts out in the form of highly energetic gamma rays, but is soon converted by interactions with the charged particles around it into slightly less energetic x-rays. These can be thought of as like particles of very energetic light, or photons. Photons interact with charged particles, in effect bouncing off them. So they have great difficulty getting out of the Sun. Each x-ray photon is bounced around between the charged particles inside the Sun like a ball in a crazy pinball machine, going forwards, backwards, up, down and sideways at random. Because of the difference in pressure between the centre and the outer part of the Sun, it is slightly easier for the photons to move outward, and they gradually leak up towards the surface.

If a photon could move in a straight line from the centre of the Sun to its surface, the journey would take 2.5 seconds, moving at the speed of light. But it actually follows such a convoluted zig-zag path that it takes, on average, 10 million years to make the journey. If the zig-zag path followed by the photon could be stretched out in a straight line, it would literally stretch for 10 million light years. Each gamma ray released in the core of the Sun is ultimately responsible for several million photons of visible light escaping from the Sun's surface. The surface of the Sun is shining today thanks to nuclear reactions that were going on at the heart of the Sun 10 million years ago, when our ancestors were still living in the trees of Africa; but we know that the nuclear reactions are still going on today thanks to our observations of solar neutrinos.

Halfway from the centre of the Sun to the surface, the density is the same as the density of water. Two-thirds of the way out, it is the same as the density of the air we breathe. About 85 per cent of the way out towards the surface, a dramatic change occurs. It is the top of what is known as the radiation zone, and the bottom of the convection zone. There, the density is only 1 per cent of the density of water, and the temperature is only 500,000 °C (900,000 °F).

This is where electrons and nuclei can begin to get together to make atoms, and instead of bouncing photons around these absorb the energetic photons from the interior. The energy absorbed in this way heats the gas in the outer layer of the Sun like a pan of water being heated from beneath on a stove. Just like the water in the pan, the top layer of the Sun undergoes convection as the hot material rises up, radiates energy at the surface of the Sun, cools and sinks (fig. 43).

The convection zone extends over about 140,000 to 100,000 km (87,000 to 62,140 miles), the outer 15 to 20 per

Fig. 43. The internal structure of the Sun.

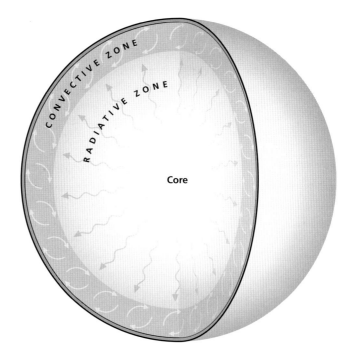

cent of the Sun, a bit less than half the distance from the Earth to the Moon. The top of the convection zone forms the visible bright surface of the Sun, called the photosphere. The temperature there is only 5800 kelvin (κ),[8] and the density is less than one-millionth of the density of water, even though the gravitational pull is 27 times the pull at the surface of the Earth. From the photosphere, light zips across space to the Earth, 149,597,870 km (92,955,810 miles) away from the centre of the Sun (148,901,605 km, 92,523,170 miles, from its surface), in just 8.3 minutes. The distance from the centre of the Sun to the centre of the Earth is just over 107 times the diameter of the Sun: almost exactly 215 solar radii.

Spots and Flares

We can watch how the Sun rotates because its surface is marked by dark spots that last typically for an interval of between 10 days and a month. The Sun rotates once every 25.38 days on average, but the higher latitudes rotate more slowly than the equator. Spots near the equator take 25 days to go round the Sun once, but at latitudes of 30 degrees

8__Remember that zero on the Kelvin scale is −273.15 degrees on the Celsius scale. Astronomers generally prefer to use κ.

north or south the rotation period is 27.5 days, and at 75 degrees it is 33 days. The average rotation period of 25.38 days actually occurs at a latitude of 15 degrees on either side of the equator. This differential rotation is possible because the Sun is a fluid object, not solid. The rotation is in the same direction as the motion of the Earth in its orbit around the Sun, so the Sun has to rotate a little bit extra to catch up with the Earth as it moves, and from Earth the average rotation of the Sun seems to take 27.27 days. The rotation flattens the Sun slightly, making the equator bulge slightly so that the solar diameter across the equator is 10 km (6.2 miles) more than the equator measured from pole to pole. The Sun differs from a perfect sphere by just 0.001 per cent.

Fig. 44. A massive sunspot (with a diameter of about 15 Earths) recorded by the spaceprobe SOHO in 2003.

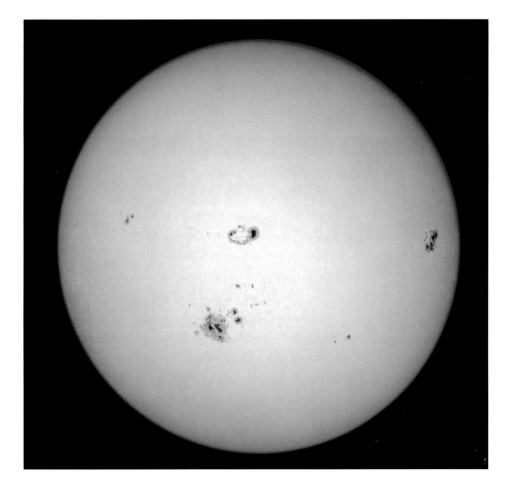

Sunspots are associated with regions of intense magnetic activity, which suppresses convection and allows the region to cool (fig. 44). Very often the spots appear in pairs, one with a north magnetic pole and the other with a south magnetic pole, like the poles of a bar magnet. This shows that magnetic field is emerging from the surface of the Sun at one spot of the pair, looping over in an arch and going back through the site of the other pole of the pair. Sunspots appear dark by contrast with the hotter, brighter regions surrounding them, but they would look quite bright against a dark background. The temperature in the middle of a sunspot is about 1500 K lower than their surroundings, but this is still around 4500 K.

An individual spot may be a small pore about 1500 km (930 miles) across, or a large, sprawling region 150,000 km (93,210 miles) across, big enough to stretch halfway from the Earth to the Moon. Spots usually form in groups, like a temporary archipelago of islands, which can extend over hundreds of millions of square kilometres. An individual spot may last for as little as a day; a typical group of sunspots lasts for less than a single rotation of the Sun (25.38 days) but

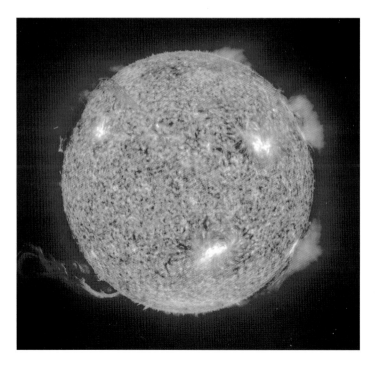

Fig. 45. *A solar flare showing a huge eruptive prominence, at temperatures of 60,000 to 80,000 K. It is much cooler than the surrounding corona, which is typically more than 1 million K.*

many last long enough to be seen twice as the Sun rotates. A major sunspot characteristically develops over an interval of about a week, then fades away over a couple of weeks.

Sunspots are also associated with solar flares, brilliant bursts of light that last for only a few hours (sometimes only about 20 minutes) and send steams of energetic charged particles, mostly electrons, out into space (*fig. 45*). A typical flare releases about one-hundred-thousandth as much energy as the entire Sun while it lasts, but the largest flares release a hundred times as much energy, about a thousandth of the entire output of the Sun. Flares typically reach heights of 5000 km to 15,000 km (3110 to 9320 miles) above the photosphere. The particles from flares interact with the Earth's magnetic field and particles high in the atmosphere to produce unusually intense aurorae; the electromagnetic disturbances they create can disrupt communications and even knock out power lines, causing widespread blackouts at high latitudes on Earth.

Flares themselves are often associated with eruptions of material from the solar surface called prominences. The gas erupted in these outbursts travels at as much as 1000 km (620 miles) a second and reaches temperatures as high as 30,000 K, often forming a looping arch above the surface of the Sun following the looping of the magnetic field. They can reach heights of hundreds of thousands of kilometres, well over twice the distance from the Earth to the Moon, with a base tens of thousands of kilometres long. But they only last for a short time. There is also a quieter family of longer-lasting prominences, which form shapes described as loops, jets, fountains, trees and filaments; they may persist for several rotations of the Sun. A typical quiet prominence may extend across 200,000 km (124,300 miles) of the Sun's surface and reach a height of 40,000 km (24,860 miles) above the surface.

The Solar Cycle

Flares, spots and other surface activity on the Sun are all associated with the Sun's magnetic field. This magnetism must be generated by the movement of electrically charged particles deep within the spinning Sun, but the exact details of how this happens are still a mystery. The strangest thing about the Sun's magnetic field is that it does not stay the same all the time, but reverses, with north and south magnetic poles swapping over, in a fairly regular cycle roughly 22 years long. This seems to be similar to the way the Earth's magnetic field reverses (see page 37), but with a much steadier, and much more regular, rhythm.

The most obvious feature of the solar cycle is a variation in the number of spots visible on its surface. The number of spots builds to a peak, dies away and then builds up to another peak over a cycle roughly 11 years long, measured peak-to-peak – the sunspot cycle (fig. 46). But the solar magnetic field reverses at the end of each sunspot cycle, so it takes two of these cycles to complete a magnetic cycle. This is sometimes known as the double sunspot cycle, but it is the true solar cycle of activity. Individual sunspot cycles may be as short as 8 years or as long as 16 years. The 'strength' of a sunspot cycle, measured in terms of the number of sunspots at the peak, also varies from cycle to cycle.

Each cycle starts with a few spots appearing at high latitudes, between about 30° and 45° from the equator in both hemispheres of the Sun. Roughly speaking, the build up to maximum activity takes about 4.5 years, and as solar activity builds up, new spots form at lower latitudes as well, closer and closer to the solar equator. Then, the activity declines over an interval of about 6.5 years, with spots no longer appearing at high latitudes, and new spots only forming at lower and lower latitudes. But the spots never form closer than about 7° to the equator. Spots linked to the next cycle of activity may already be appearing at high latitudes while there are still a few spots belonging to the previous cycle appearing at low latitudes. Maximum activity

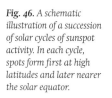

Fig. 46. A schematic illustration of a succession of solar cycles of sunspot activity. In each cycle, spots form first at high latitudes and later nearer the solar equator.

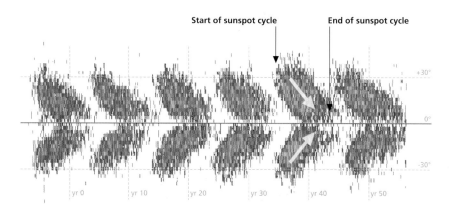

is usually confined to a single year (the period is known as solar maximum), but the quietest phase of the Sun is spread out over two or three years (solar minimum).

This pattern of activity is thought to be caused by the rotation of the Sun winding up the magnetic field beneath the surface, twisting it until something has to give and the magnetic field reverses. But this is no more than an educated guess.

Sunspots are just the most visible aspect of the overall solar cycle of activity, and there are more flares, for example, when the Sun is more active. This means that the solar wind of particles streaming outward from the Sun is more intense at solar maximum, and there seems to be a link between the changing strength of the solar cycle of activity and the weather on Earth. The length of each cycle and the strength of each cycle, both measured in terms of sunspot numbers, both vary, and there is some evidence that these variations may be linked to 'supercycles' 76 and 180 years long. There is also some evidence that when the Sun is less active overall (when there are few sunspots even at solar maximum) the Earth cools slightly.

The Warmth of the Sun

Similar cycles are hinted at in the long-term record of climatic variations on Earth. But the best evidence for this solar–terrestrial connection comes from the historical records of sunspots that were observed in the seventeenth and eighteenth centuries. These show that there were very few sunspots visible in the 70 years following 1645, which coincided (if that is the right word) with a period of cold in Europe known as the Little Ice Age. But there is no entirely satisfactory explanation of the potential link between sunspots and weather.

Some people have tried to explain the recent global warming in terms of an increase in solar activity, rather than accepting the evidence that it is largely caused by a build up of greenhouse gases in the atmosphere caused by human activities. The Sun certainly was very active in the late twentieth century. Unfortunately for this argument, though, judging by the temperature changes associated with the Little Ice Age, even if the solar–terrestrial effect is real, it is too small to account for the global warming of the twentieth century, and in the twenty-first century the warming has intensified even though solar activity has declined and ought, if anything, to be making the world cooler.

It is, of course, thanks to the Sun that the Earth is warm enough for life forms like us to exist at all. The amount of energy from the Sun crossing each square metre of a sphere centred on the Sun with the radius of the Earth's orbit is called the solar constant. It amounts to 1.37 kilowatts per square metre, and this doesn't sound a lot when you compare it with the output of a simple electric fire, typically a kilowatt. But that energy is crossing *every* square metre of the imaginary sphere around the Sun, all the time. It is as if the Sun were completely surrounded by a shell of electric fires at the distance of the Earth's orbit, all pouring their energy outward into space. Taking the radius of the Earth's orbit as 150 million km (93 million miles), in round numbers, that adds up to 283 thousand billion billion (283×10^{21}) electric fires, without even allowing for the extra 0.37 kilowatts per square metre.

One way of putting this in perspective is to imagine that, instead of the electric fires, the Sun was surrounded by a shell of ice 2.5 cm thick at the distance of the Earth's orbit, one astronomical unit. The heat from the Sun would melt all of the ice in 2 hours and 12 minutes. Now imagine that this whole shell of ice is shrunk down towards the Sun, but keeping the same amount of ice in it, so that as the radius gets smaller the thickness of the shell increases. By the time the inner surface of the ice is touching the Sun, the shell would be more than 1.5 km (0.9 miles) thick; but the prodigious output of energy from the Sun would still melt it in 2 hours and 12 minutes. And this huge output of heat has been going on for well over 4 billion years.

Actually, the solar constant isn't quite constant. The output of energy does change slightly as the Sun ages. As the nuclear fuel in the interior gets used up, the output of the Sun increases steadily, and over the lifetime of the Sun so far it has increased by about a third. This means that it was about 25 per cent cooler than it is today just after it formed. If the Sun were that cool today, the Earth would freeze. But when the Sun and the Earth were both young, the atmosphere of the Earth was rich in gases like methane and carbon dioxide that trap heat through the greenhouse effect. Because of the effect of life on the atmosphere, the greenhouse effect gradually declined as the Sun has warmed, keeping the temperature at the surface of our planet comfortable for life. This controlling influence of life on the long-term climate of the Earth is one of the main planks of Gaia theory. But almost all the greenhouse gases have now been removed from the atmosphere, and there is little scope for any further reduction. As the process of solar warming continues, in about a billion years from now the Earth will become uncomfortably hot for life. As we shall describe in Chapter Six, even more dramatic and interesting things happen to a star like the Sun at the end of its life as a main sequence star.

The Solar Wind

Because it is not a solid object, the Sun does not have a distinct surface like the Earth or the Moon; the edge we use to measure its diameter is the bottom of the region from which light escapes into space: the base of the photosphere. The photosphere is really a region several hundred kilometres thick. Below the photosphere, the Sun is opaque; above the photosphere, light escapes freely into space. The photosphere itself is slightly more transparent than the atmosphere of the Earth. Just above the photosphere, at an altitude of about 500 km (310 miles), we find the coolest region of the Sun's atmosphere, with a temperature of about 4000 K. This is cool enough to allow the formation of simple molecules, such as carbon monoxide and water vapour, which can be detected spectroscopically. But above this minimum the temperature rises considerably, probably because of the input of energy from events like solar flares. The layer from about 500 km (310 miles) to about 10,000 km (6200 miles) is called the chromosphere, and can be seen as a bright ring around the dark shadow of the Moon during a total solar eclipse. From observations made during a solar eclipse seen in 1868, spectroscopic studies revealed the presence in the chromosphere of an element unknown on Earth; it was named helium, after the Greek name for the Sun, *helios*. It was only in 1895, three decades later, that helium was isolated on Earth and found to have exactly the spectroscopic features required to explain the solar observations. The temperature in the chromosphere rises from about 4000 K at its base to as much as 100,000 K at the top.

At the top of the chromosphere there is a narrow transition region in which the temperature rises rapidly from about 100,000 K to about 1 million K. The source of the energy for this heating is not well understood, but at least some of it comes from the loops of magnetic energy associated with prominences. Above this region, the corona forms an extended outer atmosphere, which is clearly visible during a total solar eclipse, as a pearly haze around the eclipsed Sun (fig. 47). It actually shines only slightly less brightly than the Full Moon and cannot usually be seen against the glare of the Sun. It is maintained at a temperature of several million K by magnetic energy and perhaps other, as yet unidentified, sources.

The brightly visible part of the corona extends over only a few solar radii, and its appearance depends on the state of the solar cycle of activity. Because the corona merges with the outward flow of particles that forms the solar wind, and does not have a clearly defined edge, for convenience astronomers define the upper boundary of the corona as the place where the outward flow of particles is moving faster than waves can move through it. So any disturbance farther out from the Sun than this cannot feed back into the corona, because it is being carried outward faster than any ripples from the disturbance are moving in

Fig. 47. *A solar eclipse showing the corona, February 1998.*

towards the Sun. As the solar wind blows away into space, the Sun loses about a billion kilograms of matter (a million tonnes) every second, about a fifth as much as the mass it loses in the form of energy. This is equivalent to a lump of terrestrial rock 125 metres (410 feet) across being lost every second, but this is only a tiny fraction of the Sun's mass of 1.99×10^{30} kilograms.

Beyond the corona, the region in which the Sun's magnetic field dominates over the weak magnetism of interstellar space is called the heliosphere. This is like the magnetosphere of the Earth, but much larger. Just as the shape of the magnetosphere of the Earth is distorted by the influence of the solar wind, so the heliosphere is thought to be distorted by a 'cosmic wind' of tenuous material in interstellar space. The boundary of the heliosphere lies far beyond the orbits of all the planets; in the early years of the twenty-first century, observations made by the spaceprobe Voyager 1 suggested that this boundary, called the heliopause, lies between 150 AU and 160 AU from the Sun.

Most of the heliosphere, from the top of the corona to the heliopause, is filled with the solar wind. This stream of electrically charged particles consists mostly of fast-moving electrons and protons – at the distance of the Earth from the Sun the average speed of the wind is 450 km (280 miles) per second, but this varies between about 200 km per second and 900 km per second (120 to 560 miles per second) depending on how active the Sun is. The density of the solar wind near the Earth's orbit averages out at about seven protons per cubic centimetre, but this varies considerably. For comparison, there are about 20 billion billion (20×10^{18}) particles in every cubic centimetre of the air you breathe. But over billions of years even the low density of fast-moving particles in the solar wind would have eroded away much of the Earth's atmosphere, if it were not for the shielding effect of the Earth's magnetic field. It is largely thanks to the magnetosphere that you actually have a significant amount of air to breathe (fig. 48).

Because, unlike the air you breathe, the solar wind is electrically charged, it interacts with the Sun's magnetic field and stretches it into a spiral pattern produced by a combination of the outward flow of the wind and the rotation of the Sun itself. This pattern is sometimes likened to the pattern of water spraying from a rotating lawn

Fig. 48. The Earth's magnetic field. This shields us from a solar wind of particles that could be harmful to life on Earth.

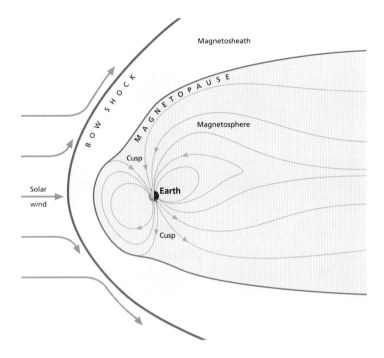

Magnetosheath

BOW SHOCK

MAGNETOPAUSE

Magnetosphere

Cusp

Solar

wind

Earth

Cusp

sprinkler. The region where the magnetic field is dominated by the solar wind in this way begins about 20 solar radii above the surface of the Sun, a tenth of the way from the Sun to the Earth, and this can be regarded as the extreme upper limit of the corona.

Variations in the strength and speed of the solar wind produce a kind of space weather, which causes problems to satellites and spaceprobes and is a potential health hazard for astronauts similar to the problems caused by nuclear radiation. At its most powerful the particles in the wind can damage solar panels and instruments on spaceprobes, while the pressure of the wind drags on satellites and speeds their re-entry into the Earth's atmosphere. Some outbursts on the Sun eject fast-moving blobs of material known as coronal mass ejections into the solar wind; these are the equivalent of space storms.

Although particles from the solar wind do cross the heliopause and become part of the cosmic wind of interstellar material, the Sun's story essentially ends at this boundary. But before we continue our journey outward to the stars, we should take a look at those tiny specks of dirt in the emptiness of space around the Sun: the planets.

Fig. 49 (facing) A neutrino 'telescope' – the Super-Kamiokande detector.

Chapter_Four

The Inner Solar System

Planets and Rocks

Although the whole pattern of stars on the sky seems to rotate around us every 24 hours as the Earth spins on its axis, the stars do not appear to move relative to one another during a human lifetime. They seem to be fixed in the same patterns night after night. However, there are a few objects that *can* be seen moving against the background of the fixed stars from day to day and year to year. The Ancient Greeks gave these objects the name 'planets', from their word for a wanderer. We now know that these wanderers are other worlds that we can see moving because, like the Earth, they are in orbit around the Sun (fig. 51). The stars do, in fact, move; but they are so far away that it takes many years of painstaking observations to see them change their positions. The planets are in our own astronomical backyard, so it is easy to see the way they move.

There is another important distinction between stars and planets. Stars shine because they are hot – kept hot by nuclear reactions going on deep in their hearts. Planets (and moons) shine only because they reflect light. In the case of the Solar System, the planets shine by reflected sunlight; planets in other systems reflect the light from their own central star.

Our astronomical backyard is called the 'Solar System' because it is dominated by the Sun – 'Sol' was the name of the Roman Sun god. The Solar System consists of the Sun itself and everything that is in orbit around the Sun, held in the grip of the Sun's gravity. In the inner part of the Solar System there are four rocky planets, more or less like the Earth and the Moon, at different distances from the Sun. Outside the orbits of these planets there is a band of rocky debris called the asteroid belt. Sometimes, lumps of rock

Fig. 50 (overleaf)
Artist's rendering of a massive asteroid belt detected around a star the same age and size as our Sun. The belt circles a faint, nearby star called HD 69830, 41 light years away in the direction of the constellation Puppis. The belt was discovered by NASA's Spitzer Space Telescope. Compared to our own Solar System's asteroid belt, this one is larger and closer to its star – it is 25 times as massive, and lies just inside an orbit equivalent to that of Venus.

from the asteroid belt fall in towards the Sun, passing close by the inner planets, or even colliding with them. Beyond the asteroid belt there are four very large planets, which are mostly made of gas, and beyond them a band of icy debris called the Kuiper Belt. The whole Solar System is surrounded by a shell of icy lumps, like an eggshell surrounding the yolk, called the Oort Cloud. Sometimes, lumps of ice from the Oort Cloud fall in towards the Sun. As they heat up, gas streams out from them to make a long, glowing tail, and they become comets. All of the planets (fig. 51) orbit around the Sun in the same direction, and all in the same plane, known as the ecliptic; it is the plane of the Sun's equator.

Fig. 51. The relative sizes of the Sun and the planets of the Solar System. These are shown to scale in this illustration. However, the distances of each planet from the Sun are not to scale.
1. Mercury
2. Venus
3. Earth
4. Mars
5. Jupiter
6. Saturn
7. Uranus
8. Neptune

Fig. 52. *The distance to a nearby star can be determined by measuring the angle across which the star seems to move, against the background of more distant stars, from opposite sides of the Earth's orbit round the Sun. This is known as parallax.*

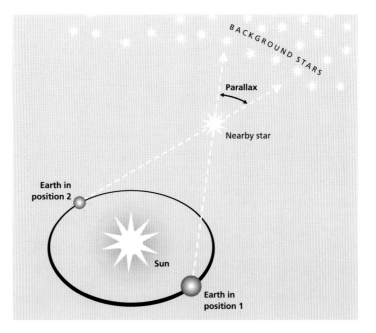

Unmanned spaceprobes have now visited every planet of the Solar System, and landed on some of them. So we no longer have to rely on observations made from Earth using telescopes to explore the Solar System (fig. 53), and we know a great deal about each of its components. So much so, that it makes sense to devote plenty of space to the Sun's family, starting with the inner Solar System.

Mercury

The closest planet to the Sun is called Mercury. Its average distance from the Sun is only 0.39 AU (39 per cent of the distance from the Sun to the Earth), so it is never very far from the Sun. It is hard to see because of the Sun's glare, although there are certain times when it can be picked out either just before the Sun rises or just after the Sun sets, depending on where it is in its orbit. The closest that orbit takes Mercury to the Sun is 46 million km (28 million miles); the farthest it gets from the Sun is 70 million km (43 million miles). This makes its orbit more elliptical (less circular) than the orbits of the other planets. Mercury takes almost exactly 88 of our days to orbit around the Sun once, so its 'year' is 88 Earth days long. But it spins very slowly on its axis. Because of the way it is locked in the Sun's gravitational grip, it rotates once every 58.6 of our days, exactly two-thirds of its year. Because of this, the Sun rises and sets very slowly on Mercury, and from a point on the surface of Mercury the Sun would be seen to move around the sky once every 176 Earth days. In that time, Mercury would have gone round the Sun twice, but turned on its axis three times. So the 'day' on Mercury is 176 of our days long, twice as long as its year.

Because Mercury is so close to the Sun, it gets very hot at noon. The Sun would look two and a half times bigger than it does from Earth, and the temperature would soar to above 450 °C (842 °F). But during the long night the surface

Fig. 53. Big Bear Solar Observatory in California, with its dome built to house a 1.6 metre (5 feet 3 inches) telescope.

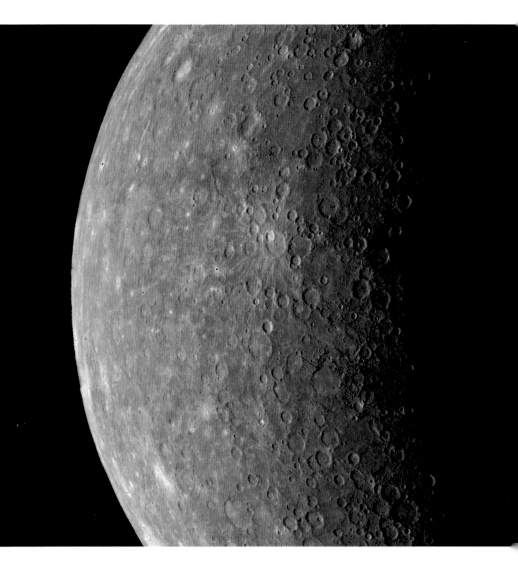

Fig. 54. *This picture of Mercury was taken at a distance of 5,380,000 km (3,340,000 miles) by the Mariner 10 spaceprobe in March 1974. The planet's diameter is about one third that of Earth.*

has plenty of time to cool off, and the temperature drops to about −180 °C (−292 °F), one of the coldest places in the Solar System even though it is one of the nearest to the Sun. The reason for the extreme difference in temperatures from one side of the planet to the other is that Mercury has no blanket of air to carry heat around it in the form of winds.

More than anything else, Mercury resembles our Moon. It was visited by a spaceprobe known as Mariner 10 in 1974 and 1975, and pictures of Mercury show a cratered surface just like the cratered surface of the Moon (*fig. 54*). It takes an expert to tell which is which in close-up images. The diameter of Mercury is only 4880 km (3030 miles), which is just 50 per cent bigger than the diameter of our Moon. But that is enough to make subtle differences between the craters on the Moon and on Mercury. Because Mercury is bigger, its gravity is stronger and the craters on Mercury are shallower than their lunar counterparts – debris could not be thrown out from the impact sites so easily. But it is quite clear that, like the lunar craters, the craters on Mercury were produced by an intense bombardment of rocks from space when the Solar System was young.

The biggest crater on Mercury is the Caloris Basin (*fig. 55*), which spans 1340 km (830 miles). It was made when an object about 150 km (90 miles) across struck the planet nearly 4 billion years ago. The impact was so powerful that it sent shock waves rippling right through the planet to

Fig. 55. Mercury's Caloris Basin in November 2001. This is the largest basin on Mercury (some 800 miles across) and is named Caloris (Greek for 'hot') because it is one of the two areas on the planet that face the Sun at perihelion.

converge on the other side of Mercury, where they have formed a jumbled hilly terrain as big as France and Germany put together.

By studying the way the gravity of Mercury affected the trajectory of Mariner 10, astronomers have calculated the mass of Mercury, which is just 5.5 per cent of the mass of the Earth, and worked out its density. This shows that it is at least 70 per cent iron and nickel with a rocky crust making up the other 30 per cent of its mass. For comparison, the Earth's iron-rich core makes up only 32 per cent of its mass; in a way, Mercury is like the core of the Earth without the surrounding material, and it also has a small magnetic field. So far, Mariner 10 is the only spaceprobe to have visited Mercury, but on 3 August 2004 NASA launched the Messenger spacecraft on a long voyage through the inner Solar System. In order to get to Mercury, Messenger first used the gravity of Earth to give it a 'slingshot' on its way in August 2005, then in October 2006 and June 2007 it made two passes by Venus. It made its first fly-by of Mercury in January 2008, a second in October that year, and will make a third in September 2009 before finally settling into orbit around Mercury in March 2011 to begin a year-long, close-up study of the planet. The Messenger cameras will be able to pick out features just 18 metres (60 feet) across, compared with the resolution of 1.6 km (1 mile) provided by Mariner 10's instruments.

Venus

As this itinerary for Messenger implies, Venus is the next planet out from the Sun after Mercury. It is far enough away from the Sun to be visible as a very bright object in the sky (the brightest apart from the Sun and the Moon) when it is in the right position in its orbit. Like Mercury, it is sometimes seen as an evening 'star' just after sunset, and sometimes as a morning 'star' just before sunrise; for this reason, the ancients thought that they were seeing two different planets.

Venus orbits the Sun once every 225 days at an average distance of 0.72 AU (108 million km, 67 million miles), nearly twice the distance of Mercury, in a very nearly circular orbit. At its closest to us, it comes within 40 million km (25 million miles) of the Earth, closer than any other planet. In

terms of its size, Venus is very nearly a twin of the Earth, with a diameter of 12,100 km (7520 miles) (just 650 km, or 400 miles, less than that of the Earth) and 82 per cent of the mass of our home planet. But the surface of Venus is very different from the surface of the Earth. It looks so bright in the sky because it is covered by a thick layer of white clouds, containing sulphuric acid, which reflect 80 per cent of the incoming sunlight and also shield the surface of the planet from our view. Astronomers only began to understand what the surface of Venus is like when they were able to send spaceprobes to visit the surface of the planet. Some have landed on the surface, surviving briefly to send back

Fig. 56. A false-colour photo-mosaic of the hemispheric view of Venus, as revealed by more than a decade of radar investigations culminating in the 1990−94 Magellan mission.

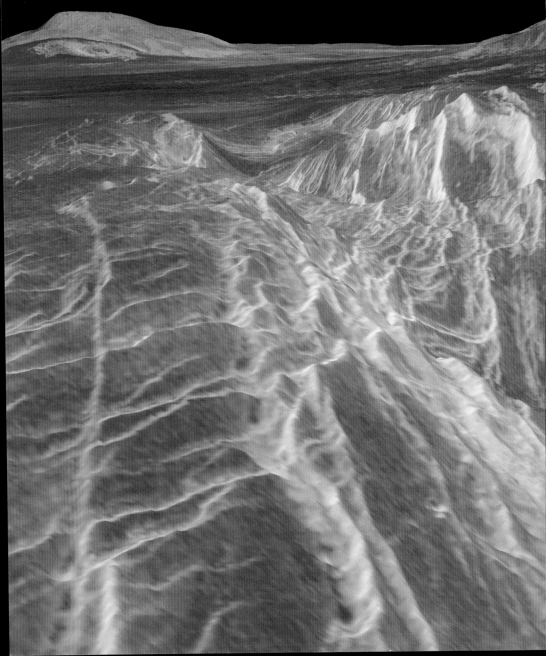

Fig. 57. *A false-colour perspective of Venus's volcanic mountains Sif (left) and Gula Mons (right).*

data to Earth before being crushed by the extremely dense atmosphere, but the most detailed information has come from radar mapping of the surface of Venus carried out from orbit around the planet. A probe called Venus Express, which is studying the atmosphere of the planet, not mapping its surface, was actually in orbit around Venus at the time Messenger flew by in June 2007.

In spite of the blanket of cloud cover that reflects away so much of the solar energy, the surface of Venus is very hot. This is because it has a very thick atmosphere, mostly made of carbon dioxide, which traps heat through the greenhouse effect. The pressure of the atmosphere at the surface of Venus is 90 times the pressure of the atmosphere at sea level on Earth, and the temperature is about 500 °C (932 °F). Because of the thick atmosphere, the temperature everywhere on Venus is about the same, regardless of whether it is day or night. But it is also dark – only about 1 per cent of the incoming sunlight penetrates to the ground.

The main reason why Venus has such a thick atmosphere is that there is no water on Venus. Because it is closer to the Sun than we are, Venus was never cool enough for liquid water to flow on its surface. On Earth, carbon dioxide has been taken out of the atmosphere and laid down in rocks rich in carbonate, such as limestone and chalk, by the action of water and life. There is almost exactly the same amount of carbon dioxide in the rocks of the Earth's crust as there is in the atmosphere of Venus. But because Venus is closer to the Sun, it was never cool enough for oceans to form, so the carbon dioxide stayed in the atmosphere.

The rotation of Venus is unique – it is the only planet in the Solar System that rotates backward, from east to west, and it does so very slowly, taking 243 Earth days to turn once on its axis. Because of the combination of the rotation of the planet and its orbital motion around the Sun, if you could see the Sun from its surface the time from one noon to the next would be 117 of our days, so there would be just under two Venusian days in each Venusian year (225 Earth days).

Although the surface of Venus is completely dry, like the surface of the Earth it has low-lying regions and higher regions. The highest mountain range on Venus, Maxwell Montes, rises 12 km (7.5 miles) above the average surface level, even more than the height of Mount Everest above sea level, just under 9 km (5.5 miles). But such mountains are much rarer on Venus than on Earth. There is a big contrast between the depth of the ocean and the heights of mountain ranges in many places on Earth, but nearly two-thirds of the surface of Venus is within 500 metres (0.3 miles) of the average planetary radius, which is equivalent to sea level on Earth. Only 15 per cent of Venus is made up of highlands and mountains, the equivalent of continents on Earth, but a third of the Earth is made up of continents, the terrestrial equivalent of the highlands of Venus.

Geologically speaking, the big difference between Venus and Earth is that there is no trace of plate tectonics and continental drift on Venus. There are volcanoes, and the higher regions are probably produced by upwelling material from the hot interior of the planet, but there is no sideways movement or subduction. This may be partly because there is no water. On Earth, water is involved in the chemical transformation of molten rock into different kinds of magma, and in effect acts as a lubricant to help tectonic plates slide over the underlying layers. Another, more important, reason is that the crust on Venus is much thicker than the crust of the Earth. This is probably because so much of the Earth's original crust was lost in the impact that led to the formation of the Moon.

This lack of continental drift seems to have had a spectacular effect on the way Venus releases heat from its interior. Just as on Earth, heat builds up inside Venus because of radioactive decay, and this has to escape somehow through the insulating layer of the crust. On Earth, this is a steady process involving volcanoes, spreading ridges and plate tectonics. But on Venus it seems to happen in violent outbursts at long intervals.

When astronomers first studied detailed radar maps of Venus, they were surprised to find that it has relatively few craters on its surface. The Moon and Mercury have been bombarded with cosmic debris since the planets formed, a bit more than 4 billion years ago, and they have craters on top of craters in some places. From counting the number of craters on Venus and comparing this with the number of craters on Mercury and the Moon, astronomers calculate that the surface of Venus has only been bombarded for about 600 million years. In other words, the surface we see today is only 600 million years old. Their explanation for

Fig. 58. A computer-simulated view of Venus based on images obtained by the Venera 13 and 14 landers in 1993. The bright feature near the centre is Ovda Regio, a mountainous region in the western portion of the great Aphrodite equatorial highland.

this is that 600 million years ago the entire surface of Venus must have been flooded with molten rock from the interior, which burst out through cracks in the old surface when the pressure became too much. With the heat released, the new surface cooled and solidified in a smooth layer over the planet. Since then, the surface has once again been like a blanket surrounding the hot core of the planet, where radioactivity will go on making it hotter and hotter until the surface cracks again. It is thought that Venus must have been 'resurfaced' in this way several times since the Solar System formed.

Mars

Moving outward through the Solar System from Venus we find the Earth–Moon double planet, which is, by definition, at a distance of 1 AU from the Sun, roughly 50 per cent farther out than Venus. About 50 per cent farther out from the Sun than the Earth–Moon system, we find the planet Mars, which has a slightly elliptical orbit taking it as close as 1.38 AU to the Sun and as far out as 1.67 AU. The closest Mars ever gets to Earth as they each move around the Sun is 55 million km (34 million miles), which happens only every 15 years. The diameter of Mars is 6790 km (4220 miles), a little more than half the diameter of the Earth, and its mass is a little more than one-tenth of the mass of the Earth. But its day is almost exactly the same length as ours – 24 hours 37 minutes, compared with our 23 hours 56 minutes. Because it is farther from the Sun, though, the year on Mars is much longer than ours, at 687 Earth days. Mars has been the most intensely studied planet, both in the days when astronomers had to rely on Earth-based telescopes and in the days of spaceprobes, so it is the planet we know most about, apart from our own.

Mars is a desert planet with a thin atmosphere where temperatures range from about − 26 °C to − 110 °C. The

Fig. 59. *A colour-composite of Mars taken in 1990, when it was mid-summer in the planet's southern hemisphere. The thin atmosphere appears relatively clear of dust over most of the planet. However, a thick canopy of clouds obscures the icy north polar regions, located near the centre of the bright bluish clouds at the top of the image.*

atmospheric pressure at the surface of Mars is the same as the pressure in our atmosphere 35 km (21.7 miles) above sea level. Although one of the main reasons why people are fascinated by Mars is the hope that it might be a home for life, if James Lovelock's Gaia idea is correct the search for life on Mars is doomed because we know it has a stable atmosphere in which none of the chemical processes associated with life have left their mark. It is often known as the 'Red Planet', because of its distinctive reddish-orange colour; this is caused by the reddish dust that covers the surface of the planet and is often carried high into the atmosphere in dust storms.

In spite of the extreme conditions on Mars, there is water there, locked up in ice at the poles, and possibly frozen underground. Like the other planets of the inner Solar System, Mars is scarred by craters caused by impacts, and the patterns of debris around many of these craters show the kinds of features associated with flowing water, presumably from ice that was melted by the impact and froze again soon after. There are also many channels and other features that seem to have been produced by flowing water, but they are ancient features. This is taken as a sign that Mars used to be much warmer, more than 3.5 billion years ago when the planet was young, and that it had liquid water flowing across its surface in the form of rivers, and perhaps even oceans. Most probably, when Mars was young

Fig. 60. Map of the surface of Mars.

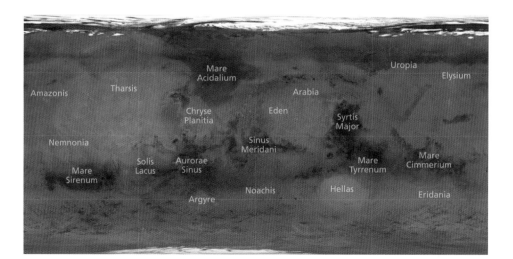

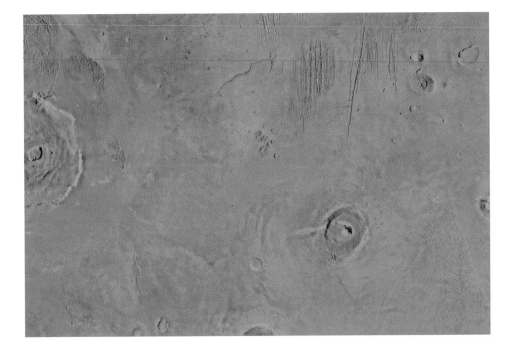

it had a thick enough carbon dioxide atmosphere for the surface to be kept warmer than the melting point of ice by the greenhouse effect, even though it is so far from the Sun. But because Mars is so small, its gravitational pull was not strong enough to hold on to this atmosphere, and because it has no significant magnetic field there was nothing to shield it from the influence of the solar wind, which would have scoured away its atmosphere over billions of years.

So most of the atmosphere escaped into space as the planet settled into a deep freeze. Today, the surface of the planet is a red, frozen desert.

The atmosphere could not be renewed by volcanic activity, because Mars is too small to produce the internal heat required to maintain enough such activity. It would have cooled off very quickly after it formed. So, like Venus, but for a different reason, there is no continental drift on Mars. It does, though, have the largest volcano in the Solar System, the Nix Olympica, which rises just over 26 km (16 miles) above the Martian equivalent of sea level and covers an area the size of Arizona. There are other volcanoes almost as impressive on Mars, and they are probably still growing, but very slowly.

Fig. 61. Tharsis region of Mars. Three of the four largest shield volcanoes on Mars – Olympus, Ascraeus, and Pavonis Montes – lie within the Tharsis quadrangle, together with several smaller volcanoes.

Some of the largest of these volcanoes are associated with a huge uplift in the Martian crust, near the equator, known as Tharsis (*fig. 62*). The Tharsis uplift has raised a region of crust about 4000 km (2490 miles) across up by as much as 10 km (6.2 miles) compared with its surroundings. This region lies on the boundary between two very different Martian hemispheres. The northern hemisphere of Mars is mostly made up of low-lying plains, lower than the Martian equivalent of sea level. By contrast, most of the southern hemisphere is covered by highlands, between 1 and 4 km (2.5 miles) above 'sea level', which are heavily cratered, like Mercury and the Moon. The cratering shows that the surface of Mars is as old as the surface of Mercury or the Moon, and has not been reworked like the surface of Venus. But nobody knows why there is such a difference between the two hemispheres of the planet; perhaps it suffered a huge impact billions of years ago.

The kind of object that causes the cratering on the inner planets can be seen orbiting around Mars today, in the form of its two tiny moons, Phobos and Deimos (*fig. 63*). These are so small and hard to see that they were only discovered in 1877, by the astronomer Asaph Hall (1829–1907), working at the US Naval Observatory in Washington, DC. They are both irregularly shaped lumps of rock – Deimos is about 15 km by 12 km by 11 km (9.3 by 7.5 by 6.8 miles) in size, and Phobos is 27 km by 21 km by 19 km (16.7 by 13 by 11.8 miles). Phobos orbits around Mars three times a day (once every 0.319 Earth days), at an altitude just 6000 km (3730 miles) above the surface (a distance of 9380 km, or 5830 miles, from the centre of Mars); Deimos orbits at a distance of 23,460 km (14,580 miles) once every 1.26 Earth days.

The Asteroid Belt

Phobos and Deimos might be leftover fragments of the material from which Mars formed, but most likely they are lumps of rock from the general debris left over from the formation of the Solar System, which wandered too close to Mars and got captured in its gravitational grip. Much more of this debris is found in a band of cosmic rubble, the asteroid belt, which lies between about 1.7 AU and 4 AU from the Sun (*fig. 64*). This region of the Solar System contains more than a million individual asteroids a kilometre or more

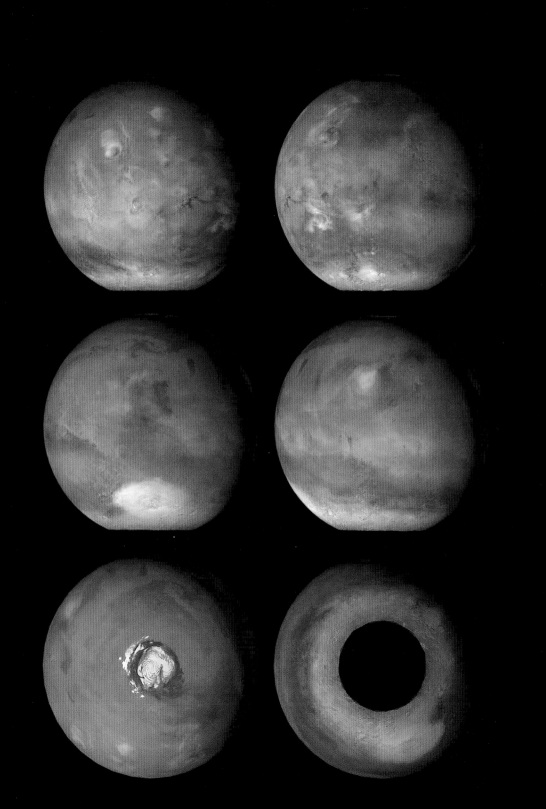

across, similar to the moons of Mars, and countless numbers of smaller members ranging down to pieces the size of a grain of sand. More than 100 thousand asteroids have been photographed, but only about 10 thousand have been studied and given numbers and (in many cases) names. Several spaceprobes have passed through the asteroid belt on their way to the outer Solar System and photographed different asteroids; in February 2001 one probe, called NEAR Shoemaker, even landed on the surface of an asteroid known as 433 Eros.

The largest asteroid, Ceres, orbits the Sun at a distance of 2.8 AU, with an orbital period of 4.6 years. It has a diameter of 933 km (580 miles) and contains more than a quarter of all the mass of the asteroid belt; only 10 of the asteroids are bigger than 250 km across (160 miles), and only 120 are bigger than 125 km (80 miles). But Ceres has a dark surface and is not the brightest asteroid, in spite of its size. That honour goes to the third-largest asteroid, Vesta, which is only 500 km (310 miles) in diameter but has such a light surface that at times it can be seen with the naked eye from Earth. The total amount of matter in the asteroid belt only adds up to about 15 per cent of the mass of our Moon. Asteroids are also known as minor planets; depending on their position in the asteroid belt, they take between 3 and 6 years to orbit around the Sun, mostly in roughly circular orbits (fig. 64).

Fig. 63. *Asteroid Gaspra (top) and the moons of Mars: Deimos (lower left) and Phobos (lower right). Ranging in diameter from 17 to 27 km (10.6 to 16.8 miles), all three have irregular shapes, due to past catastrophic conditions. However, their surfaces appear remarkably different, most likely because of very different impact histories. The Phobos and Deimos images were obtained by the Viking Orbiter spacecraft in 1977, the Gaspra image by the Galileo spaceprobe in 1991.*

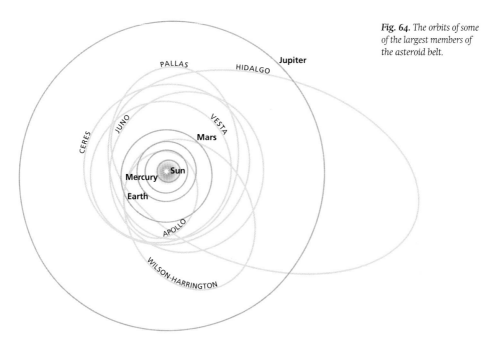

Fig. 64. *The orbits of some of the largest members of the asteroid belt.*

A few of these objects, about 5 per cent, have elliptical orbits that take them much closer to the Sun, crossing the Earth's orbit on the way. Some of these pieces of cosmic debris collide with the Earth and other inner planets. Small pieces of material burn up in the atmosphere of the Earth as meteors, or 'shooting stars'; larger pieces hit the surface and, if they are large enough, make impact craters. They are then called meteorites. Several thousand small meteorites reach the surface of the Earth each year, without doing any significant damage. But at the other extreme, as we have mentioned, there have been large impacts from space like those that may have been responsible for the death of the dinosaurs and other extinctions of life on Earth. Even these occasional large impacts are nothing, though, compared with the battering that the inner planets received when the Solar System was young.

Making Rocky Planets

The Sun and Solar System formed from a cloud of material in space that collapsed under the influence of gravity. This cloud consisted mainly of hydrogen and helium gas, which, as we have seen, settled into the centre of the collapsing

cloud to form a star, the Sun. But there was also a relatively small amount of dust in the cloud, no more than 2
per cent, in the form of particles as fine as the solid
particles in a cloud of cigarette smoke. As the Sun became
hot, leftover gas around the star was blown away, but
because the collapsing cloud was rotating – everything
in the Universe rotates to some extent – the solid material left outside the collapsing core of the cloud did not
fall into the star but settled into a disc around the rotating
young Sun. Such dusty discs, known as proto-planetary
discs or PPDs, have been photographed around young stars
today (fig. 65).

Fig. 65. *An artist's
impression of a flared
proto-planetary disc,
similar to the one detected
around the 2.5 solar mass
star HD 97048.*

Samples obtained from meteorites show that in the dusty disc around the young Sun there were also tiny globules (known as chondrules), which had melted at hot spots in the disc, at temperatures between 1,200 and 1,600 °C (2192 to 2912 °F), and then solidified quickly in space. Nobody knows what caused the heating. The tiny particles in the inner part of the disc around the young Sun, dust and chondrules, were able to begin to stick together because they were all moving the same way around the Sun. There were very few head-on collisions, just more gentle bumps when a grain overtook another grain and nudged it from behind. This process gradually built up fluffy balls of stuff that began to clump together under the influence of gravity. It would only take about a hundred thousand years, a tiny amount of time compared with the present age of the Solar System, 4.6 billion years, to build up objects a kilometre or so in size. These would continue to join together to form larger and larger objects as the remaining gas was swept out of the inner Solar System by the heat of the Sun. The kilometre-sized objects from which planets formed are known as planetisimals.

Computer simulations show that within 20,000 years there would be hundreds of objects the size of our Moon. Within about a million years from the time when the original cloud of gas and dust started to collapse, there would have been twenty or so objects with sizes ranging from that of the Moon to that of Mars moving around the Sun within a distance of about 2.5 AU, roughly the distance from the Sun to the asteroid belt. These were accompanied by large numbers of smaller objects, some of which have survived to become some of the minor planets of the asteroid belt itself. The bigger objects in the inner Solar System swept up most of the primordial debris, however, and some of them collided with one another and merged, leaving just the four rocky planets we see in the inner Solar System today – Mercury, Venus, Earth and Mars – plus our Moon. The planets would reach roughly their present sizes within 10 million years, and then keep on sweeping up material in the inner Solar System for another 100 million years or so.

One of the most spectacular of these collisions was the one in which the Earth–Moon system formed. But there must have been at least one equally dramatic event when two Mars-sized objects collided and broke each other apart, producing some of the other objects now found in the asteroid belt. Some of these minor planets show clear signs of having been melted at some time in the ancient past, and one explanation for this is that they were part of a planet-sized object that got hot enough for melting to take place before it was broken apart. This fits in with the idea that there used to be a lot more material in the asteroid belt and the rest of the inner Solar System, but that it has been cleaned up as the orbits of the asteroids were disturbed by the gravitational pull of the planets while the Solar System was young, sending meteorites plunging on to the Earth and the other inner planets.

You can see the evidence for this just by looking at the Moon. Within a little more than 10 million years after the Sun started to shine, the Earth and the other inner planets had settled down as hot but cooling balls of rock. By a 100 million years after the birth of the Sun they had solid crusts. For the next few hundred million years, the surfaces of all these planets (and the Moon) were bombarded by meteorites. The craters on airless Mercury and the Moon show the effects of this battering most clearly. Mars is also heavily cratered, although the traces of craters have been softened by erosion over billions of years, even in the thin Martian atmosphere. On Earth, the primordial surface has mostly been turned over by the processes of plate tectonics and continental drift, while Venus has suffered at least one cataclysmic overturn, so there is no record of the early bombardment, but it must have happened here as well. The battering ended – or rather, thinned out – about 4 billion years ago. Astonishingly, within another 200 million years, by 3.8 billion years ago, there was life on Earth, revealed by the presence of fossils in rocks that old. Those rocks also show that there was abundant liquid water on Earth at that time, even though the planet had only recently, by the standards of astronomical time, been a red-hot ball of rock.

The sudden emergence of our planet as a suitable home for life can be explained, as we shall see in Chapter Nine, by looking farther out across the Solar System, beyond the asteroid belt, where the processes of planet formation were different, and where the debris left over from the formation of the Solar System was in the form of mountains of ice, not lumps of rock.

Fig. 66. *The Rocky Mesas of Nilosyrtis Mensae region of Mars in May 2007. Clay minerals have been detected in this region by ESO's Mars Express Orbiter and from the Mars Reconnaissance Orbiter.*

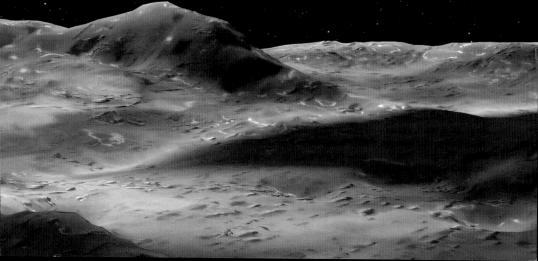

Chapter_Five

The Outer Solar System

Gas Giants and Ice Mountains

The outer part of the Solar System has a similar structure to the inner part of the Solar System, but on a larger scale. In the inner Solar System, there are four relatively small, rocky planets and an outer ring of rocky debris, the asteroid belt. In the outer Solar System, there are four large, mostly gaseous planets (the fact that it is the same number seems to be just a coincidence) and an outer ring of icy debris, the Kuiper Belt. Beyond everything else, literally halfway to the nearest star, the entire Solar System is surrounded by a spherical shell, known as the Oort Cloud, containing much more icy debris. Everything is held in orbit by the Sun's gravity.

The gas giant planets – Jupiter, Saturn, Uranus and Neptune – are different from the rocky planets because they formed in a different way and they are in a different part of the Solar System. It used to be thought that both kinds of planet formed in the same way, as small pieces of material stuck together to make larger objects. On that picture, the first part of a giant planet to form would have been a rocky core, and this would attract gas to itself because of its gravitational pull. The idea was that the outer planets grew bigger simply because there was more gas in the outer part of the Solar System, because the heat of the young Sun was feebler there and had not blown the gas away, so big planets formed there. But this picture is wrong. Simple calculations show that a lump of ice or rock ten or twelve times the mass of the Earth in the orbit Jupiter is in today could indeed grow to the present size of Jupiter by attracting material from a gas cloud in this way, but it would take a very long time. In the case of Uranus and Neptune, the two outermost planets, this 'bottom up' process would take

Fig. 67 (overleaf)
An artist's concept of the Pluto system from the surface of one of its possible small moons. Pluto is the large disc at centre, right. Charon, the system's only confirmed moon, is the smaller disc to the right of Pluto. The other candidate small moon is the bright dot on Pluto's far left.

Fig. 68. Jupiter showing its stripy cloud pattern and the Giant Red Spot. The spot appears blue in this image because of the filters used to enhance the stripy pattern of the clouds.

longer than the present age of the Solar System. So we can be sure that the gas giants did not form from a bottom up process in their present orbits.

The alternative is that they formed by a 'top down' process. If there were clumps of greater than average density in the swirling disc of material around the young Sun, as there surely must have been, it would be easy for the largest of these lumps to collapse to make large balls of gas like the outer planets. Such balls of gas could form anywhere in the disc, close to the Sun or farther out. In the past decade or so, astronomers have indeed found more than a hundred planetary systems around other stars, in most cases with a giant planet relatively close to its parent star – a kind of 'hot Jupiter'. These are the easiest planets to find outside our Solar System, so these planetary systems are not necessarily typical, but they do prove that giant planets can form anywhere in the disc around a young star. Nobody knows for sure why all the gas giants in our Solar System are far from the Sun, but here is the best modern thinking on the subject.

Making Giant Planets

These discoveries have encouraged the development of a new explanation of how the outer planets of our own Solar System got to be the way they are today. This model is still incomplete, and more details need to be worked out, but it is probably roughly right. It takes on board evidence from lunar samples obtained on the Apollo missions more than 30 years ago that the Moon, and presumably the rest of the inner Solar System, suffered an intense bombardment by meteorites when the Solar System was about 700 million years old, just before life emerged on Earth. This is known as the late heavy bombardment, or LHB.

According to the new model, all four of the gas giant planets formed close to one another, surrounded by a swirling mass of planetesimals made of ice and rock. Because of the way the young planets tugged on each other gravitationally, Jupiter slowly moved closer to the Sun, but the other three giant planets moved outward. Some 700 million years after the Solar System formed, these changes produced a pattern in which Saturn was in an orbit with a period exactly twice as long as Jupiter's orbit, and this produced a rhythmic gravitational tug on the smaller objects in the outer Solar System. This rapidly moved Uranus and Neptune farther out from the Sun. In particular, the orbit of Neptune suddenly doubled in size, sending the planet into the inner part of the then much larger Kuiper Belt. Its presence there disturbed the orbits of many of the smaller objects. The gravitational influence of Neptune shook up the material there so that some of the smaller objects moved faster and some of them slower. This sent huge numbers of planetisimals in towards the Sun, where they collided with the inner planets in the late heavy bombardment, and others out into the depths of space. Some of the pieces of debris were also captured by the giant planets as moons, circling them like planets orbiting the Sun, as Galileo discovered four centuries ago.

This also explains why there are no planets like Earth, Mercury, Venus or Mars in the outer Solar System. The pieces from which such planets might have formed either got hoovered up by the giants, or flung away into space or the inner Solar System, or became moons of the giant planets. Whatever the details of how the gas giant planets and their moons got to be where they are today, though, what matters now is that they are there, and that we know more about them than any previous generation.

Jupiter

Moving outward from the asteroid belt, the first planet we encounter is Jupiter, by far the largest of all the planets in the Solar System. It contains 317 times as much mass as the Earth, 0.1 per cent as much mass as the Sun, and has a diameter eleven times the diameter of the Earth. Jupiter takes 11.86 years to orbit the Sun once, at a distance of 5.2 AU. The closest Jupiter ever gets to Earth is 590 million km (370 million miles) away, but because of its size and the highly reflective clouds in its atmosphere, at its brightest as seen from Earth it is brighter than any of the other planets except Venus, and brighter than every star in the night sky except Sirius. The mass of Jupiter is more than twice as much as the mass of all the other planets put together, but it is chiefly composed of hydrogen and a smaller amount of helium, the two lightest elements; it is more like a failed star (see Chapter Six) than an Earth-like planet. Although it is so big, Jupiter has a low density over-all, just 30 per cent more than the density of water, because most of its mass is concentrated near the centre of the planet. Jupiter rotates so rapidly, once in just under ten hours, that it has a distinct bulge, making its diameter 9276 km (5760 miles) longer measured across its widest point than from pole to pole. The equatorial diameter is 142,984 km (88,850 miles), but the polar diameter is only 133,708 km (83,090 miles).

Because Jupiter is so big, it has a powerful gravitational influence and holds on to many moons, like a miniature Solar System. New moons of Jupiter are still being discov-ered. At the time of writing, the count had reached the mid-sixties, but most of these are only irregular lumps of cosmic debris a few kilometres across. By far the most important are the four largest, the 'Galilean satellites', so named in honour of their discoverer. These are Io, Europa, Ganymede and Callisto (fig. 69). Among the many spacecraft that have visited Jupiter, the orbiter Galileo released a probe that went into the atmosphere on 7 December 1995, surv-iving for 57.6 minutes under a parachute before being crushed by the pressure 150 km (93 miles) below the top of the visible cloud surface.

The visible 'surface' of Jupiter is just the top layer of clouds in its gaseous atmosphere. These form a series of coloured

Fig. 69. Jupiter's four large and diverse 'Galilean' satel-lites, from left to right: Io, Europa, Ganymede and Callisto. As seen by the New Horizons spacecraft during its fly-by of Jupiter in late February 2007. The images have been scaled to rep-resent the true relative sizes of the four moons and are arranged in their order from Jupiter.

horizontal stripes, bright where hot gas is rising by convection and darker where cold gas is descending. 'Hot' and 'cold' are relative terms, and the clouds high above the bright stripes are full of frozen crystals of ammonia. The average temperature at the top of the clouds of Jupiter is -150 °C (-238 °F). The colours of the clouds are mostly different shades of yellow, brown, orange and red, but can be purple, because of the different chemical compounds they contain, including hydrocarbons (compounds containing hydrogen and carbon). The bands of cloud move at different speeds around the planet, up to 400 km (250 miles) per hour, creating eddies like giant whirlpools along their edges.

One of the most prominent features of Jupiter is a huge eddy called the Great Red Spot (fig. 68). It was first observed by Robert Hooke in 1664, although it was probably around long before then. Although its appearance has varied, it has always been there since its discovery. The spot is caused by a spiralling upward flow of gas like a huge hurricane, bulging up 8 km (4.9 miles) above the surrounding clouds. The spot is about 14,000 km (8700 miles) long and 40,000 km (24,850 miles) wide, so it could swallow three planets the size of the Earth side by side. A smaller red spot formed in the early twenty-first century when several smaller white eddies merged.

Even though it is very far from the Sun, Jupiter has such an active atmosphere because it does not depend on heat from the distant Sun for energy. Jupiter actually radiates out into space twice as much heat as it receives from the Sun, because it has an internal store of warmth left over from the time it formed. Ever since Jupiter formed, it has been shrinking under its own weight, converting gravitational energy into heat. It was probably about twice its present size when it formed, and it is still shrinking by about two centimetres every year. Convection caused by this heating from within drives the winds on Jupiter and features such as the Great Red Spot.

There may be a solid rocky core to Jupiter, but if so it is tiny compared with the mass of the planet – no more than twelve times the mass of the Earth. Below the cloud layers we can see, which form a layer only about 50 km (31 miles) deep, the temperature is similar to that on Earth, and there are clouds containing water vapour. Deeper still, about

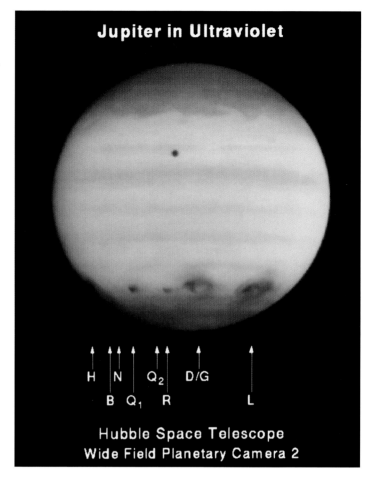

Fig. 70. Ultraviolet image
of Jupiter taken by the
Hubble Space Telescope.
The image shows Jupiter's
atmosphere at a wavelength
of 255 nm after many
impacts by fragments of
comet Shoemaker-Levy 9.

1000 km (620 miles) below the visible surface, the pressure
is so extreme that the hydrogen–helium mixture is compressed into a liquid, forming an ocean 20,000 km (12,430
miles) deep with a core where the pressure is 3 million
times greater than the pressure of the atmosphere at the
surface of the Earth. The hydrogen in the heart of Jupiter is
so compressed that it becomes a metal, at a temperature of
several tens of thousands of degrees Celsius, and currents
stirred up by convection in the metallic hydrogen and the
rotation of the planet produce a magnetic field fourteen
times stronger than the Earth's magnetic field, extending
out into space out a hundred times the radius of Jupiter. If
we could see this magnetic field, from Earth it would look
more than twice as big as the Moon on the sky.

The Galilean Moons

Electrically charged particles trapped in this magnetic field interact powerfully with Jupiter's moon, Io, which is closer to the cloud tops of the planet than the Moon is to Earth; the radius of its orbit is 421,700 km (262,030 miles) from the centre of Jupiter, just 350,208 km (217,610 miles) above the equatorial clouds (fig. 71). Io is slightly bigger than our Moon, with a diameter of 3643 km (2263 miles), and completes one orbit around Jupiter every $42\frac{1}{2}$ hours. To the surprise of astronomers, when Voyager 1 passed by and photographed Io in 1979 it discovered that the moon is the most volcanically active object in the entire Solar System. The volcanoes spew out liquid sulphur, which solidifies across the surface of Io in red, orange and yellow patches.

Io is too small to have retained a hot core of molten material in the way the Earth has, but there are two processes that are thought to keep its interior molten today. The first is a kind of gravitational squeezing caused by the tidal

Fig. 71. Composite image of Io, the most volcanic body in the Solar System, as viewed by NASA's Galileo spaceprobe in 1996. The smallest features discernible are 2.5 km (1.5 miles) across. There are rugged mountains several kilometres high, layered materials forming plateaus, and many irregular depressions (volcanic calderas). The side shown is that facing away from Jupiter, with north to the top.

influence of Jupiter and the other Galilean moons, which regularly squeeze and release their grip on the moon. The other possibility is that as Io orbits in Jupiter's intense magnetic field, electric currents are generated in its interior, heating it like electricity passing through the bar of an electric fire. So much sulphur has erupted from Io that some has escaped and coated the small moon Amalthea, which is an irregular lump of rock just 200 km (124 miles) across that orbits even closer to Jupiter.

The next Galilean satellite, Europa, is the smallest, with a diameter of just 3122 km (1939 miles). Europa is covered by a shell of ice, marked with cracks, but there is probably a rocky core beneath the ice crust. Its orbital radius is 671,034 km (416,960 miles), and it takes 3.55 days to travel round Jupiter once. Close-up views show that in many places the fractured surface of Europa is broken up into slabs of ice several kilometres across which resemble the pack ice seen in the polar seas of Earth during the spring thaw. These 'plates' of ice have clearly jostled against one another, and must be lubricated by a layer of water or slushy ice beneath them.

Beyond Europa, we come to Ganymede, the largest moon in the Solar System, with a diameter of 5262 km (3269 miles).

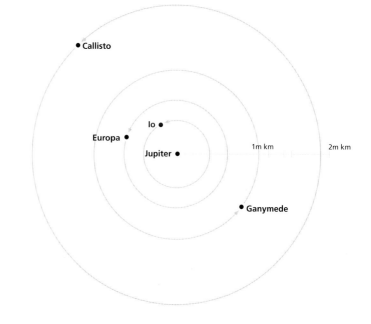

Fig. 72. Jupiter and its four 'Galilean' satellites are like a miniature Solar System.

It has an orbit with a radius of just over a million km (621,400 miles) and takes 7.15 days to complete each circuit of Jupiter. Ganymede is even larger than the planet Mercury, and is thought to have a rocky, iron-rich core surrounded by an icy outer shell.

The fourth Galilean satellite, Callisto, has a more uniform structure, with rock and ice mixed together like a snowball studded with stones. The cause of the difference is that Callisto never got hot enough for its interior to melt and allow heavy material to settle into the centre. Callisto has a diameter of 4821 km (2995 miles), orbits at a distance of nearly 1.9 million km (1.1 million miles) from the centre of Jupiter, and takes 16.69 days to go round the planet once.

Callisto's surface is smothered in impact craters, like those on the surfaces of the Moon and Mercury, caused by meteorites striking its surface. The largest, called Valhalla, is 300 km (186 miles) across (*fig. 73*). Ganymede also has a

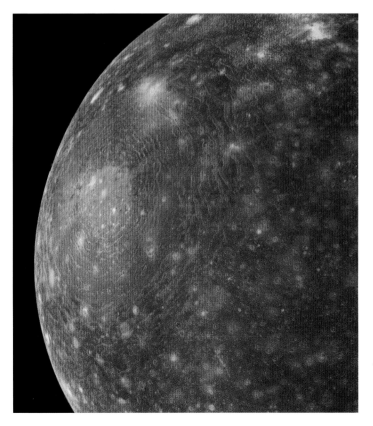

Fig. 73. Callisto, revealed by the Voyager cameras in 1979 to be a heavily cratered and hence geologically inactive world. The prominent old impact feature Valhalla has a central bright spot about 600 km (370 miles) across, probably representing the original impact basin.

cratered surface, but is not as heavily cratered as Callisto. Recent craters on both moons show up as bright patches where the underlying ice has been exposed and has not had time to get covered in dust.

One interesting feature of the three inner Galilean moons is that their orbits form a pattern known as a resonance (see page 116). Because of the gravitational interaction between the satellites, for every four orbits that Io makes around Jupiter, Europa makes two orbits and Ganymede makes one. When Europa and Ganymede are closest together, Io is always on the opposite side of Jupiter. The effect of all this is to stretch each of the orbits slightly, making them more elliptical. At the same time, the effect of Jupiter's gravity is to try to make the orbits more circular. This conflict is the source of the stretching and squeezing that helps to keep the heart of Io hot; the other two moons experience a similar squeezing, but because they are farther from Jupiter the effect is smaller and they do not get so hot inside. But this may explain why Europa is warm enough inside for liquid water to lie beneath its icy surface, and why Ganymede once had a molten core but Callisto did not.

As well as its large retinue of moons Jupiter has a faint ring system, discovered by the spaceprobe Voyager 1 in 1979. These rings had not been noticed from Earth because they do not reflect much sunlight. Unlike the bright rings of Saturn (the next planet out from the Sun after Jupiter), which are made of ice, the rings of Jupiter seem to be made of dust ejected from its moons by asteroid impacts.

Saturn

Saturn's rings make the planet the most spectacular astronomical object visible through a small telescope. When the angle between the Sun, the rings and the Earth is just right, they actually reflect more light to us than the planet itself does. But at times we see them edge on, and they essentially become invisible from Earth. Even Galileo's telescopes showed him that there was something odd about the appearance of Saturn, as if there were a pair of ears on the disc of the planet, but he never saw the rings in all their glory.

Saturn takes 29.46 years to orbit the Sun once, at a distance of 9.5 AU. It is much smaller than Jupiter, but with an equatorial diameter of 120,536 km (74,900 miles) still much bigger than any of the other planets. Saturn's polar diameter is only 108,728 km (67,560 miles), some 10 per cent less than its equatorial diameter, so it is even more flattened than Jupiter. As well as being the second-largest planet in the Solar System, it has the second-fastest rotation rate, once every 10 hours 47 minutes, and like Jupiter it is made largely of hydrogen with some helium. Overall, it is a smaller version of Jupiter.

But there is one special feature of Saturn, unique among the planets of our Solar System. Because its mass, 95 times the mass of the Earth, is less than a third of the mass of Jupiter,

Fig. 74. *Saturn. Astronomers combined ultraviolet images of Saturn's southern polar region with visible-light images of the planet and its rings to make this picture. The auroral display appears blue because of the glow of ultraviolet light. In reality, the aurora would appear red to an observer on Saturn because of the presence of glowing hydrogen in the atmosphere.*

the overall density of Saturn is only 70 per cent of the density of water. If you had an ocean big enough, Saturn would float in it. This low density tells us that even in relation to its smaller size Saturn does not have such a large, dense central core as Jupiter; the low density also explains why the equatorial bulge is so pronounced, even though Saturn spins more slowly than Jupiter. This also fits the observations of Saturn's magnetic field, which is stronger than the Earth's but weaker than Jupiter's.

Fig. 75. Saturn's moons Mimas, Enceladus and Dione viewed by the Hubble Space Telescope with the rings around Saturn. The planet's rings are tilted nearly edge-on toward the Sun, an event that occurs once every 15 years.

The stripes on Saturn are less pronounced than those of Jupiter, and there is no equivalent of the Great Red Spot. But the winds on Saturn blow even faster than those on Jupiter, up to 1800 km (1120 miles) per hour. Like Jupiter, Saturn has an internal source of heat, and radiates out into space about two and a half times as much energy as it receives from the Sun – but because it is much farther away

from the Sun than Jupiter is, this is less than the energy received by Jupiter. Once again, the energy comes from a slow shrinking of the planet.

Rings and Moons

From Earth, the rings of Saturn look like a solid disc around the planet, but in the 1850s James Clerk Maxwell (1831–79) realised that a solid disc would be broken up by tidal forces in the gravitational field of Saturn, so the disc must actually be made up of many small objects, like tiny moons, orbiting Saturn. Spaceprobes such as the Voyager series sent back pictures from Saturn showing thousands of narrow ringlets separated by gaps, looking superficially like the grooves in an old vinyl LP, with individual lumps of ice and rock following each other around in each ringlet. The rings extend from 6630 km to 120,700 km (4120 to 75,000 miles) above Saturn's equator, across a total diameter of 240,000 km (149,130 miles), but are no more than 100 metres thick, so in order to represent them properly an LP would have to be 5 km (3.1 miles) across. Individual particles in the rings range from the size of sand grains up to the size of a small car. The rings are not stable in the long term – over millions of years – so must have formed relatively recently, perhaps when a moon came too close to Saturn and was ripped apart by tidal forces.

Like Jupiter, Saturn has a large number of moons, most of which are small lumps of debris of little interest to anyone except the experts. Some of these moons interact with the rings to make the gaps. But the largest of Saturn's moons, Titan, is in many ways the most interesting moon in the Solar System (fig. 76). Titan orbits Saturn at a distance of 1.2 million km (745,650 miles) once every 15.9 days. It has a diameter of 5150 km (3200 miles), 40 per cent of the diameter of the Earth, making it bigger than Mercury (although it is less than half as massive as Mercury) but slightly smaller than Ganymede. Judging by its overall density, Titan must be made of a 50:50 mixture of ice and rock, similar to Ganymede and Callisto, probably in layers of icy material surrounding a rocky core. More importantly, however, it has a dense atmosphere, unlike any other moon in the Solar System. The atmosphere is 98 per cent nitrogen and nearly 1.5 per cent methane, producing a pressure at the surface of Titan 50 per cent greater than the pressure at sea level on Earth. Because the atmosphere is so thick and the pull of gravity is so weak at the surface of Titan, human beings on the surface would be able to fly by flapping wings attached to their arms.

Titan is more like a planet than what we usually think of as a moon, and has been described as being like a miniature frozen Earth. If it had been closer to the Sun, perhaps it would have developed an Earth-like atmosphere, and possibly provided a home for life. As it is, until recently very little could be learned about Titan from Earth, or even from fly-by spaceprobes. Orange clouds high in the atmosphere obscure the surface of the moon, where the temperature is a chilly $-180\ °$C $(-292\ °$F$)$. But our understanding of Titan took a spectacular leap forward in January 2005, when the Huygens spaceprobe landed on its surface.

Fig. 76. *Saturn's moon Titan observed by radar. The dark patches are interpreted as lakes filled with liquid methane. More than 75 dark patches were seen by Cassini in 2006, ranging from 3 km to more than 70 km (43 miles) across.*

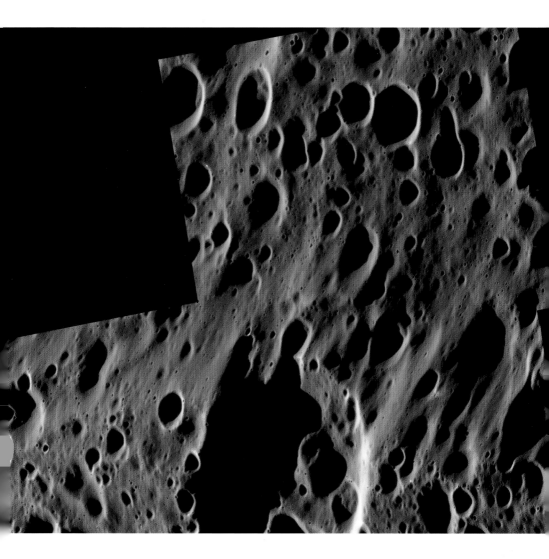

The Huygens probe was released from its parent space-craft, Cassini, which flew past Titan and remained in orbit around Saturn. Huygens spent 2 hours 28 minutes descending by parachute through the atmosphere of Titan, then sent back data from the surface for another 1 hour 10 minutes until Cassini passed out of range of its transmitter. Cassini relayed the data back to Earth. Combined with data from radar mapping and other instruments carried on Cassini, these observations opened up a new world to terrestrial astronomers. The haze in Titan's atmosphere

Fig. 77. Part of the surface of Saturn's moon Iapetus. This view shows a vast range of crater sizes in the dark terrain of the leading hemisphere of the moon. A few small bright spots indicate fresh craters where impacts have punched through the thin blanket of dark material to the cleaner ice beneath.

cleared at an altitude of 40 km (24.8 miles), so Huygens was able to take pictures of the surface as it descended, as well as after landing.

There is now evidence for lakes and seas of liquid methane (or possibly ethane) on Titan, fed by rain made of droplets of methane, and deserts striped by dunes. The seas are found near the poles, where the moon is coldest. The largest sea is nearly as big as the Caspian Sea on Earth. The surface is relatively smooth and crater-free, with few great mountains and large areas without hills more than a few tens of metres high, but there is at least one mountain range, about 150 km (93 miles) long, 30 km (18.6 miles) wide and rising 1.5 km (0.9 miles) above the surface. The few craters that have been found seem to have been filled in by wind-blown material. Huygens landed on a flat plain with a surface like sand but made of ice grains and covered with pebbles that are thought to be made of frozen water. Many of the surface features of the moon seem to have been formed by flowing fluids, but it is impossible to say when this happened.

Uranus

Saturn has one last distinction. It is the most distant planet that was known to observers in ancient times – indeed, up until the eighteenth century. The next planet out from the Sun, Uranus, was only discovered with the aid of a telescope, by William Herschel (1738–1822), on 13 March 1781, although at first he thought it was a comet; in fact it can sometimes be seen by the naked eye when the sky is very dark. Herschel was carrying out a systematic survey of the sky when he discovered Uranus, and had not been expecting to find a 'new' planet. Uranus has a slightly elliptical orbit taking it as close as 18.3 AU to the Sun and as far out as 20.1 AU, completing one orbit in just over 84 years. It weighs in with just 8.7 times the mass of the Earth, and has an equatorial diameter of 51,118 km (31,760 miles) and a polar diameter of 49,946 km (31,040 miles), nearly four times that of the Earth. Although less than half the size of Saturn, it still counts as a gas giant, although some astronomers prefer to refer to Uranus and Neptune as ice giants. Alone among the planets of the Solar System, though, Uranus lies on its side, with its equator almost at right

Fig. 78. Uranus surrounded by its rings and some of the moons. At the infrared wavelengths at which this observation was made, the disc of Uranus appears relatively dark. At the same time, the icy material in the rings reflects the sunlight and appears comparatively bright. Seven of the moons of Uranus can be identified. Of these, Titania and Oberon are the brightest. The much smaller and fainter Puck and Portia are barely visible here.

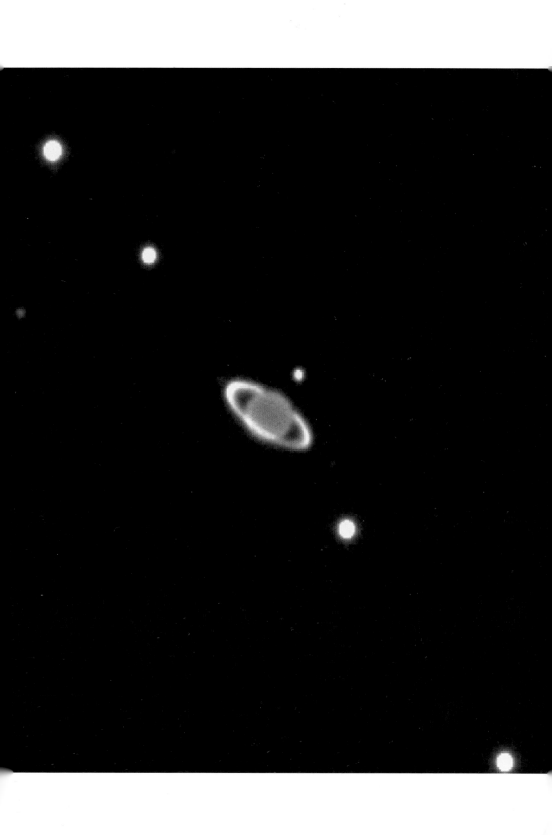

angles to an imaginary line joining the planet to the Sun; nobody knows why this is, but it may be that Uranus was struck by a large object long ago and literally knocked sideways. It takes 17 hours 14 minutes to turn once on its tilted axis.

Like Jupiter and Saturn, Uranus has a ring system, which was suspected by Herschel but only firmly discovered in 1977. The rings are very narrow compared with those of Saturn, less than 100 km (62 miles) wide and separated by gaps of hundreds or thousands of kilometres. There are five moons visible from Earth, and the usual flock of smaller satellites associated with a gas giant planet. All the moons and the rings move around the equator of the planet, even though it is lying on its side, so they must have formed after Uranus got its tilt. The visible surface of Uranus is much blander than the visible surfaces of Jupiter and Saturn, an almost uniform green with hardly any interesting features. In recent years, there have been some signs of activity in its clouds as it heads towards it closest point in its orbit round the Sun. The atmosphere seems to be mostly hydrogen and helium, but with a mixture of methane, surrounding a rocky core.

Neptune

The last known planet in the Solar System (and almost certainly really the last) is Neptune, at a distance of 30 AU from the Sun. Neptune takes 165 years to go round the Sun once, and turns on its axis every 16 hours 7 minutes. It is very nearly spherical, with a diameter of 49,000 km (30,450 miles), 3.8 times that of the Earth. This makes it almost a twin of Uranus, even down to its greenish tinge, except that it is upright in its orbit and weighs in at 17 times the mass of the Earth, nearly twice the mass of Uranus. Most of this mass is concentrated in the core of the planet.

Unlike Uranus, Neptune was not found by accident. In the nineteenth century astronomers monitoring the movement of Uranus across the sky noticed that the orbit of the newly discovered planet was not following exactly the expected path, and realised that it was being tugged on by a planet even farther out from the Sun. John Couch Adams (1819–1892), in England, and Urbain Le Verrier (1811–1877), in France, independently calculated where this unseen planet must be, and following up these predictions Neptune was found by Johann Gottfried Galle (1812–1910), working at the Berlin Observatory, on 23 September 1846. Curiously, though, drawings made by Galileo show that he saw Neptune through his telescopes at least twice, on 28 December 1612 and 27 January 1613, each time mistaking it for a star.

Neptune's mostly bland visible surface did have one interesting feature, a Great Dark Spot, photographed by Voyager 2 in August 1989 (*fig. 79*), which has since disappeared. It also has the fastest winds in the Solar System, reaching speeds of up to 2100 km (1300 miles) per hour, and two unusual satellites among its family of eight large moons and an unknown number of lesser satellites. The outermost large moon, Nereid, is only 300 km (190 miles) across and is in a very elliptical orbit that takes it out to 9.7 million km (6 million miles) from Neptune and in to 1.4 million km (869,920 miles) every 360 days. The next large moon in, Triton, is going backwards around Neptune, from east to west, and is in an unstable orbit that is gradually contracting because of tidal forces. Clearly, something has happened to disturb the

Fig. 79. *Neptune as viewed by Voyager 2 in August 1989. Bright, wispy 'cirrus type' clouds are seen overlying the Great Dark Spot at its southern (lower) margin and over its northwest (upper left) boundary. Increasing detail in global banding and in the south polar region can also be seen; a smaller dark spot at high southern latitudes is dimly visible near the limb at lower left.*

9__Named in honour of the astronomer Gerard Kuiper (1905–1973), although he did not discover it.

outer moons, but we don't know what. Triton is 2700 km (1680 miles) across, more than three-quarters the size of our Moon, and the only one of Neptune's moons big enough for gravity to have made it spherical; but it is destined to be broken up by Neptune's gravity when it gets too close to the planet in a few tens of millions of years. This will produce a ring system far more impressive than Neptune's present thin rings, and we may be seeing here the early stages of the kind of process that formed Saturn's rings.

The Kuiper Belt

Only a short distance beyond the orbit of Neptune, the Kuiper Belt[9] begins. The belt proper extends in a fat disc, or torus, from about 30 AU to about 50 AU from the Sun, but there are other icy objects both within the belt and beyond its outer edge. As we have explained, it is composed of planetisimals left over from the formation of the Solar System, like the asteroid belt, but because it is so far from the Sun these objects mostly contain various kinds of ice, in particular frozen methane, ammonia and water. Astronomers knew the belt must exist even before they identified any of these objects, from studying the orbits of so-called short-period comets, like Halley's Comet, which come from that region of the Solar System. The first objects identified as members of the Kuiper Belt were seen in 1992; more than a thousand such objects are now known, and it is estimated that there must be at least 70,000 Kuiper Belt objects, or KBOs, with diameters larger than 1 km (0.6 miles), with perhaps half of them having diameters larger than 100 km (62 miles). But the total mass in the Kuiper Belt is no more than the equivalent of about 40 times the mass of the Earth. The strange orbit of Triton may be explained if it is a KBO that has been captured by Neptune. The first KBO to be identified was given the reference QB1, and has never been formally named, so, with the fondness of astronomers for making up names, similar objects to QB1 are sometimes called cubewanos (try saying it out loud). Other KBOs similar to Pluto have been dubbed plutinos.

Pluto itself used to be regarded as a planet, but is now recognised as one of the largest members of the Kuiper Belt. It was discovered in 1930 and has an elliptical orbit that takes it as close as 29.7 AU to the Sun (closer than Neptune) and as far out as 49.3 AU. Pluto has one large moon, Charon, with half Pluto's diameter, and two smaller moons, and it goes round the Sun once every 248 years (*fig. 80*). It has only 0.2 per cent as much mass as the Earth and a diameter just over 2300 km (1430 miles), making it smaller and less massive than any of the eight major planets. After a long debate about the correct status of Pluto, the situation was resolved in 2005 with the discovery of Eris, a 'trans-Neptunian object' orbiting beyond the Kuiper Belt, which has a diameter of at least 2400 km (1490 miles) – larger than Pluto. Pluto is also smaller than Triton, which may once have been a KBO. In 2007, the mass of Eris was determined as 27 per cent bigger than Pluto's mass. So Pluto isn't even the biggest or most massive trans-Neptunian object. Pluto and Eris are now formally classified as dwarf planets. Eris, which has one tiny moon, orbits between 98 AU and 38 AU from the Sun, taking 557 years to complete one orbit. It is a member of a class of 'scattered belt objects', thought to have been members of the Kuiper Belt that have been displaced outwards by gravitational effects. It has one moon, and as of the summer of 2007 was the most distant observed member of the Solar System (*fig. 80*). But even these objects are not at the edge of the Solar System.

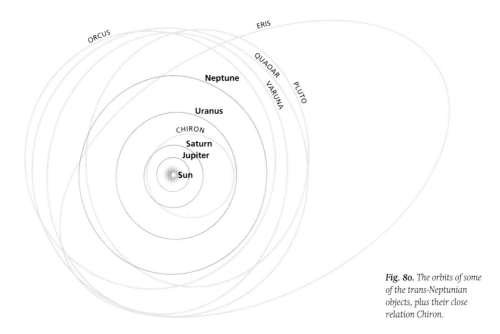

Fig. 80. *The orbits of some of the trans-Neptunian objects, plus their close relation Chiron.*

A Cloud of Comets

Just as the study of short-period comets revealed the presence of the Kuiper Belt, so the study of what are known as long-period comets reveals that there must be a reservoir of icy objects even farther out from the Sun. This is known as the Oort Cloud.

Comets are familiar, if only from photographs, as bright objects with long 'tails' of glowing gas that pass across the sky for a short time and then fade away. They are actually lumps of icy debris, tens or hundreds of kilometres across, that come from the outer part of the Solar System and pass through the inner Solar System before swinging around the Sun and going back out again. Their tails are produced by gas and dust streaming out from the core of the comet as it gets hot during its approach to the Sun. Short-period comets, like Halley's Comet, travel in orbits that extend more or less across the region of the Solar System occupied by planets, and return for a close pass by the Sun every few years or decades. Although their orbits may shift slightly because of the gravitational influence of the planets, especially Jupiter, they are easily identifiable as the same objects coming round again and again. Encke's Comet, for example, moves between 0.34 and 4.08 AU from the Sun, in an orbit with a period of 3.3 years; Halley's Comet orbits between 0.6 AU and 35 AU, with a period of some 76 years. It clearly comes from the Kuiper Belt, whereas Encke's Comet does not, or at least not recently.

But there are some comets with highly extended orbits, which come into the inner Solar System from far beyond the orbit of Neptune before swinging around the Sun and heading back out into space. Their orbits are so long that it must take them million of years to travel once around the Sun, and many of them are in fact first-time visitors to the inner Solar System. Some of them have such long, thin orbits that observations from Earth are inadequate to work out their periods at all. More than 500 long-period comets have been identified since the time of Edmond Halley, some 300 years ago, and a few more are discovered each year. One recent long-period comet to visit our part of the Solar System was Hyakutake, in 1996 (fig. 81). As comets go, Hyakutake was only average in terms of its actual brightness, but because it passed within 0.1 AU of the Earth, one-tenth

Fig. 81. Comet Hyakutake crosses the Big Dipper in March 1996.

Fig. 82. *Artist's impression of the Oort Cloud of comets that surrounds the Sun.*

of the distance between us and the Sun, it made a spectacular sight in the night sky.

Nearly half of the long-period comets coming in on such orbits are given a gravitational nudge by Jupiter that sends them out of the Solar System altogether, and the rest will gradually (over hundreds of millions of years) be slowed down into smaller orbits and become short-period comets. Yet four or five new comets come into the inner Solar System each year, even though it is more than 4 billion years since the Solar System formed. There must be a huge reservoir of comets, far beyond the orbit of Neptune, so large that it is able to supply this steady trickle of comets for all that time. What's more, because long-period comets come from all directions, not just along the plane in which the planets lie, this reservoir must form a spherical cloud around the Solar System, several light years from the Sun. The Estonian Ernst Öpik (1893–1985) calculated the possible size and location of this cloud in the 1930s, and the calculations were refined in the 1950s by the Dutch astronomer Jan Oort (1900–1992), after whom the cloud is named. Most long-period visitors come from this vast spherical shell around the Solar System, at a distance of 50,000 to 100,000 AU from the Sun (about 0.75 to 1.5 light years, over a third of the way to the nearest star). It is thought that between the Kuiper Belt and the Oort Cloud proper there is a region of space where icy objects are distributed, fanning out, like a cross-section through a trumpet, to link the two systems.

Comets in the Oort Cloud are tens of millions of kilometres apart, and drift slowly in almost circular orbits, tenuously held in the Sun's gravitational grip. Passing stars or interactions with clouds of interstellar gas can easily disturb their orbits, which is why a few fall in towards the Sun each year. The total mass of material in the Oort Cloud is estimated to be at most about a few tens of times the mass of the Earth, similar to the mass of the Kuiper Belt, distributed among several thousand billion separate objects. At these huge distances from the Sun, the temperature of a typical icy lump will be only about $-270\ °C$, three or four degrees above the absolute zero of temperature, which is $-273\ °C$. This shell of frozen material really does mark the boundary of the Solar System. Beyond that, the stars.

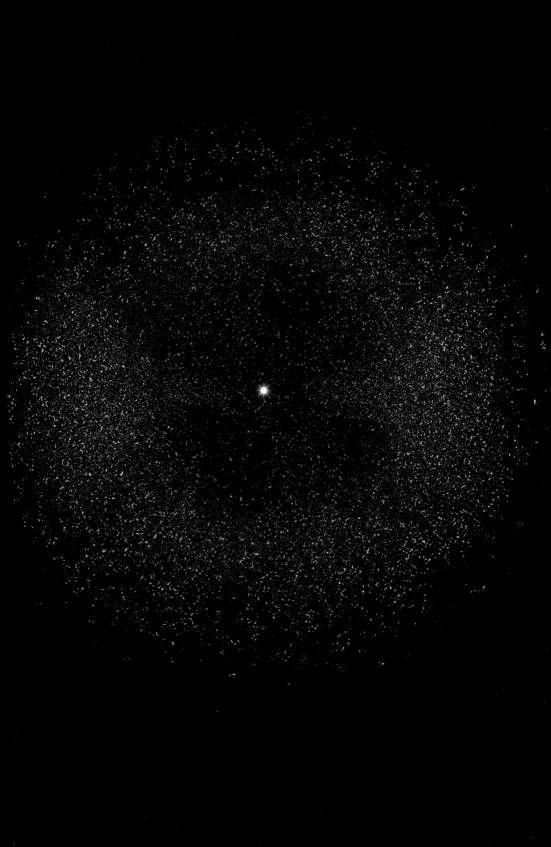

Chapter_Six

Stars

On a dark, clear, moonless night, far away from any bright city lights, the stars shine like jewels in the sky, seemingly in such profusion that they have become a byword in literature for an abundance of something. An expression like 'as numerous as the stars in the sky' instantly conveys the impression of a mighty host. But this tells us more about the way people perceive the world about them than it does about the number of stars in the sky. In fact, over the whole sky only a few thousand individual stars are visible to the naked eye. No more than half of these can possibly be visible from any one place on Earth at any one time, since one half of our planet experiences day while the other half experiences night. Once allowance is made for the difficulty of seeing stars low on the horizon, where the atmosphere is hazy and there are trees and buildings obscuring the view, under the best possible viewing conditions, without any artificial aids you would see fewer than two thousand stars spread over the night sky.

Since the time of Galileo, astronomers have been using telescopes to study much fainter stars than we can see with our naked eyes, and since the middle of the nineteenth century they have been able to photograph the genuine profusion of stars visible from Earth with such artificial aids. Millions of individual stars have now been identified, in the sense that their positions on the sky have been recorded in catalogues, and it is estimated that hundreds of millions of stars have been photographed in surveys of the sky, with their images preserved in archives but not yet catalogued. Although only a small proportion of these known stars have been studied in detail, that is still enough for astronomers to be able to determine what different kinds of star there are, how big and how hot they are, and how

Fig. 83 (overleaf) The night sky shows the gas giant Jupiter (the bright planet at the centre) as well as the stars and cosmic dust clouds of the Milky Way hanging over the southern horizon in the early morning hours as seen from Stagecoach, Colorado.

Fig. 84. The Pleiades, a reflection nebula associated with a cluster of young stars. The cluster is about 400 light years from Earth in the northern constellation of Taurus. Seven of its brightest stars, the 'Seven Sisters', are visible to the naked eye.

10__One parsec is just over 3.26 light years. It is a unit of distance used by astronomers, originally based on parallax measurements.

they work. As we mentioned in the Introduction, the key pieces of information are the distance to a star, its brightness, and its colour (which tells us how hot it is) and other details of its spectrum (which tell us what it is made of). One other crucial ingredient is its mass. Fortunately very many stars occur in binary pairs, orbiting one another the way the Earth and the Moon orbit one another. By analysing details of the orbital motion of such stars, and knowing the law of gravity, astronomers can work out the masses of these stars. With enough masses determined in this way, they have been able to discover relationships between, for example, the colour, brightness and mass of certain kinds of star, so that they can make at least an educated guess at a star's mass just by analysing its other properties.

All of this is complicated by the large distances to even the nearest stars. The nearest star system to our Sun (actually a binary pair of stars) is Alpha Centauri, at a distance of 1.33 parsecs, just 4.3 light years.[10] Other near neighbours are 61 Cygni, at a distance of 3.4 parsecs, and Alpha Lyrae, at a distance of 8.3 parsecs. These are typical of the distances between neighbouring stars. There are slightly fewer than 500 stars within 22 parsecs (just over 70 light years) from the Sun, and the distance to the nearest large cluster of stars, the Hyades Cluster, is about 45 parsecs, or roughly 150 light years. The Hyades Cluster contains a couple of hundred stars spread over a volume of space some 3.5 parsecs (12 light years) across. It is no wonder the stars look so faint from the Earth. These are the kinds of distances astronomers have to contend with when they try to analyse the nature of stars by studying the light from them – light that has taken decades or centuries to reach us.

Red Giants and White Dwarfs

The most important aspects of the relationship between the different properties of the stars are summed up in a kind of graph known as a Hertzsprung–Russell (H-R) diagram (*fig. 85*), named after the Dane Ejnar Hertzsprung (1873–1967) and the American Henry Norris Russell (1877–1957), who each independently discovered its features in the second decade of the twentieth century. The diagram plots the actual brightness of a star, known as its absolute magni-

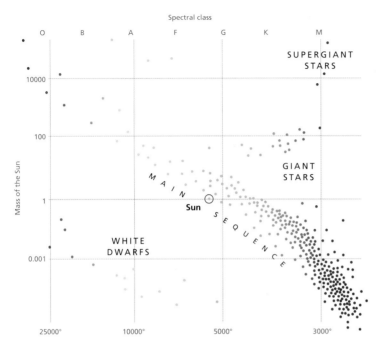

Fig. 85. *A schematic representation of the Hertzsprung–Russell diagram. One important point to notice is that our Sun is very much an average star.*

tude, against its colour, also known as its spectral class, which is a measure of its temperature. Most stars lie on a band that stretches from the top left of the diagram, corresponding to hot, bright stars, to the bottom right of the diagram, corresponding to cool, faint stars. This is known as the main sequence.

Stars on the main sequence of the H-R diagram are stars like the Sun, which are in the prime of their lives, steadily converting hydrogen into helium in their interiors. More massive stars lie near the top of the main sequence, while less massive stars lie lower down the main sequence; the Sun is an average star in the sense that it is about halfway along the main sequence, although at least 90 per cent of all stars are less massive than the Sun. Stars in the bottom right of the H-R diagram are faint, cool and red, with surface temperatures below 3,500 K. Stars in the top left of the H-R diagram are bright, hot and blue-white, with surface temperatures above 25,000 K. The most massive stars on the main sequence are about 50 times as massive as the Sun, and perhaps 20 times its diameter. The smallest have about a tenth of a solar mass of material, ten times more than the mass of Jupiter.

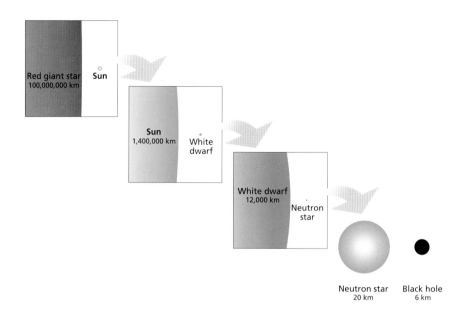

Fig. 86. Relative sizes of different kinds of star.

Above the main sequence, in the top right of the diagram, there is a more or less horizontal band of cool but bright, massive stars that are known, for reasons that will become clear, as red giants; this is the 'red giant branch'. They are bright, even though they are cool, because they are very big. In the bottom left of the diagram, below the main sequence, there are stars that are hot but faint, and must therefore be quite small; they are known as white dwarf stars.

An individual star will move around the H-R diagram as it ages, starting out just above the main sequence, moving on to the main sequence as it begins to burn hydrogen into helium, later becoming a red giant, and finally settling down as a white dwarf. All of this takes more time for less massive stars, because they do not need to burn their fuel so vigorously to hold themselves up against their own weight; it takes less time for more massive stars, which need to put out a lot of energy to resist the inward pull of gravity. But there has not been anything like enough time since the invention of the telescope to study the entire life cycle of a single star. The typical life story of a star has been worked out by studying many different stars at many different stages of their lives, just as the typical life story of a tree could be worked out by studying many different trees in a

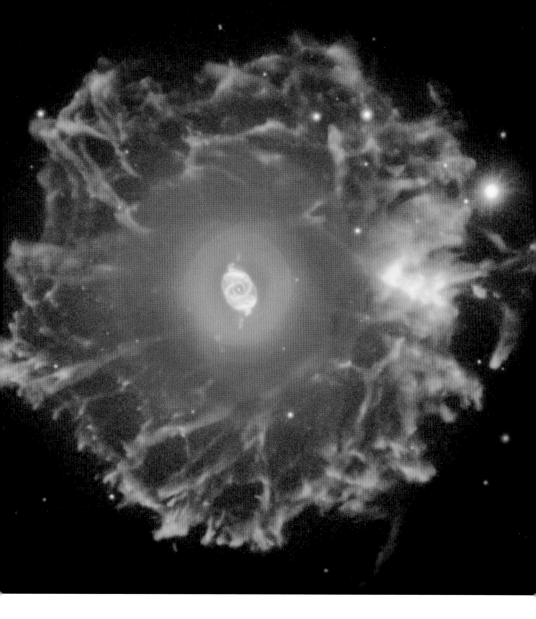

forest all at different stages of their lives. Another invaluable insight comes from studying clusters of stars; the stars in a cluster are all at roughly the same distance from us and all formed at the same time. This means that the differences we see between them must be a result of differences in their other properties, such as their mass. It all helps to establish the life story of a typical star, which begins when a cloud of gas and dust in space begins to collapse under its own weight, as a result of large-scale disturbances that will be described in the next chapter.

Fig 87. *An enormous but extremely faint halo of gaseous material surrounds the Cat's Eye Nebula and is over three light years across. Several planetary nebulae been found to have halos like this one, probably formed of material ejected during earlier active episodes in the star's evolution.*

Fig 88. *Star trails around the North Pole made by a long-exposure photograph: in the middle of the picture is the North Celestial Pole (NCP), at the centre of all the star trail arcs. The very short bright trail near the NCP was made by the star Polaris, commonly known as the North Star.*

Making Stars

Computer simulations, combined with observations of star-forming regions in space, tell us that the core on which a new star forms always starts out with about the same mass, roughly one-tenth of one per cent of the mass of the Sun, regardless of how big the eventual star will be. These pre-stellar cores are able to collapse under their own weight, getting hotter all the time as gravitational energy is converted into heat. When the centre of the core becomes dense enough and hot enough, which happens when the core is about the same size as the Sun, this internal heat produces a pressure that holds the core up and prevents any further collapse. It takes about ten years for the core to build up a mass of about one per cent of the mass of the Sun. The rest of the mass that will make up the final star falls on to the core more slowly, by accretion, and the mass of the final star depends on how much material there is in the cloud around the core, not on how big the core was in the first place.

Nuclear fusion begins when the mass of the core has reached about a fifth of a solar mass, and from then on the star has its own internal source of heat, regardless of whether it continues to accrete matter and benefit from the release of gravitational energy. The radius of the star does not increase very much during this process since the main effect of the accretion is to increase the density of the star. The accretion process ends when the young star is big enough and bright enough for its radiation, including its stellar wind, to blow away any remaining gas around it. This leaves a dusty disc of material orbiting around the young star, traces of cosmic dirt from which planets can form. At this stage of its life, a star with the same mass as our Sun will have a diameter about four times that of the Sun today, and will gradually shrink as it settles down to its life on the main sequence. The whole process takes no more than a few tens of millions of years.

But stars do not form in isolation. Any cloud of material that tries to settle down on to a single core will spin faster and faster as it shrinks, like a spinning ice skater pulling in their arms, until it is torn apart by centrifugal effects. If two or more stars form together, however, the rotation is converted into the orbital motion of the stars around one

Fig. 89. A spectacular panorama of the star-forming region in the centre of the Orion Nebula.

another. Typically, out of every 100 newly born star systems 40 are triples and 60 are binaries. Many of the triples eject one of their members to leave a binary system, and some of the binaries get torn apart by interactions with other stars, so the overall ratio ends up as 25 triples for every 65 binaries and 35 single stars. The fact that we live on a planet orbiting a single star may seem surprising statistically, but this is probably because planets like the Earth can only exist in stable orbits around single stars. In more complicated systems, the competing gravitational forces would move planets around in complicated orbits causing extreme variations in temperature that would make it impossible for intelligent life to evolve.

The Life of a Star

Once a star gets on to the main sequence, how long it stays there depends entirely on its mass. All main sequence stars have roughly the same core temperature, between 15 million and 20 million K, because they are all burning hydrogen into helium. But some burn it faster than others. A lighter star can release energy slowly and still hold itself up against its own weight; a more massive star has to burn more of its nuclear fuel each second to prevent gravitational collapse. So even though they start out with much more fuel to burn, the biggest stars burn out fastest.

The Sun is roughly halfway up the main sequence, and about halfway through its total lifetime as a main sequence star, which is likely to be a little more than 10 billion years. Astronomers have used our understanding of the physics of what is going on inside stars, combined with observations of stars at different stages of their lives, to find out how long stars with different masses stay on the main sequence. Near the middle of the main sequence, a rough rule of thumb is that the lifetime of a star is proportional to one divided by the cube of the star's mass. So a star with twice the Sun's mass stays on the main sequence for one-eighth as long as the Sun, and a star with half the Sun's mass stays on the main sequence for eight times as long as the Sun. More precisely, a star 15 times more massive than the Sun, for example, will have a main sequence life of only 10 million years, and one with three solar masses will stay on the main sequence for only 250 million years. A star with 70 per cent of the Sun's mass, by contrast, would have a main sequence lifetime of more than 20 billion years, twice that of the Sun. Since this is greater than the age of the Universe, no stars that small have actually left the main sequence yet.

A star leaves the main sequence when it is no longer able to burn hydrogen into helium in its heart. When this happens, the star begins to shrink, and this releases gravitational energy that makes it hotter in its core. At a temperature of about 100 million κ, another set of nuclear reactions can occur, fusing nuclei of helium (alpha particles) three at a time to make nuclei of carbon. The star settles down into another stable state, but its life as a helium-burning star only lasts for about 10 per cent as long as its life on the main sequence.

During this phase of its life, the extra heat from the core makes the outer layers of the star expand dramatically, and it becomes a red giant. When the Sun becomes a red giant, its outer layers will expand until it reaches beyond the orbit of Venus. The Sun will be a red giant for about a billion years; but, contrary to some accounts, it will not swell up beyond the size of the Earth's present orbit, because it will have lost a lot of mass by the time it reaches that stage of its life, as material from the expanding atmosphere gets blown away into space. Other things being equal, as the Sun loses mass its gravity weakens so that the Earth might be expected to drift out to about where the orbit of Mars is today, and be safe. Unfortunately for anyone still around on Earth, however, tidal effects are likely to overcome this effect and provide a drag on the Earth that will send it spiralling inwards, rather than outwards, so that it will not escape the fiery doom.[11]

The Sun will actually undergo two phases of expansion, the first when the core shrinks and becomes hot enough to start fusing helium into carbon. As the core contracts and gets hotter, the extra heat makes the outer layers of the star expand; but once it settles down into steady helium burning, the core cools a little and this allows the outer layers to shrink. When all the helium is used up (which will take a few hundred million years in the case of the Sun), the same sort of thing happens again. The core shrinks and gets even hotter. If a star still has as much mass then as the Sun has today, at this time of its life the outer layers will swell up so much that the star will become a red supergiant. For a

11__This is a new discovery, made in 2007 by astronomers Peter Schröder and Robert Smith.

while, it will be able to stabilise itself with an inert carbon core surrounded by a shell in which hydrogen is being converted into helium. Even after allowing for mass loss, in the first phase of expansion the Sun will extend to a radius of nearly 170 million km (105 million miles), beyond the present orbit of the Earth. So much mass will be lost, as much as 20 per cent of its original mass by the time the Sun reaches the second phase of expansion, that it will never become a true supergiant. In the second phase it will expand only a little more than in its first giant phase, to a bit more than 170 million km (105 million miles).

Planets farther out from the Sun than the Earth will not be swallowed up as the Sun expands. There really will be an outward drift for all of these planets, so Mars and the rest will be safe. Mercury and Venus, though, are, like the Earth, doomed. Mercury is so close to the Sun that it will be swallowed up long before the Sun reaches its maximum size, and although without tidal effects the orbit of Venus would expand to a radius of about 134 million km (83 million miles) by the time the Sun becomes a red giant for the first time, even if that happened it would still be orbiting within the Sun's atmosphere and get dragged inward to its fiery fate.

Stardeath

For a star with the same mass as the Sun, there is no other source of energy once the helium in the core has been used up and the last vestiges of hydrogen have been burnt up in the shell around the core. After puffing its outer layers away into space, the remaining material will settle down into a cooling lump, about the size of the Earth but still with more than half of the original mass of the star. It will have become a white dwarf, with a core rich in carbon (a stellar cinder), surrounded by a shell of helium and a thin hydrogen atmosphere. The material blown away from a star in the later stages of its life can form a glowing cloud of gas, known as a planetary nebula. Planetary nebulae are among the most beautiful objects in the Universe, but they have nothing to do with planets; they got their name because a lot of them look like little round blobs, vaguely similar to the appearance of planets, through small telescopes (fig. 90).

Stars with more than about four times as much mass as the Sun can, however, go through further phases of shrinking, getting hotter inside, and triggering further phases of nuclear fusion, making heavier and heavier elements. These reactions can build up all of the elements up to iron and related nuclei such as those of cobalt and nickel (the 'iron group'). A sufficiently massive star may end up with several different layers of material, arranged like onion skins around its core, with each different layer of material containing different nuclei undergoing different fusion reactions. But each stage of nuclear burning is over quicker than the one before. While all this activity is going on, the outward appearance of the star changes, and it may pass through phases of variable brightness, including the variations associated with a family of stars know as Cepheid variables (see *fig. 106*, page 174). What happens next depends on how massive the star is.

The critical factor is whether the remnant left behind once nuclear burning stops has more than 1.4 times the mass of our Sun, a number known as the Chandrasekhar limit, after the astronomer Subrahmanyan Chandrasekhar (1910–1995) who first realised its importance. If a stellar remnant has less mass than this, it can settle down quietly into a white dwarf. But if it has more mass than the Chandrasekhar limit, once nuclear burning stops, it will collapse even further, down into a ball only a few kilo-

Fig. 90. This is a portion of the Veil Nebula. The entire structure spans about 3 degrees on the sky, corresponding to about six times the diameter of the Moon seen from Earth.

metres across, known as a neutron star. This has about as much material as there is in a star like the Sun, compressed into the volume of a large mountain on Earth. A thimbleful of neutron star material contains as much matter as there is in the bodies of all the people alive on Earth today – each cubic centimetre contains about 100 million tonnes of matter (fig. 86).

We know this happens, because neutron stars have been detected. Many of them have strong magnetic fields and spin very fast, some of them hundreds of times every second. This produces beams of radio emission, which sweep around like the beam of a radio lighthouse, and can be detected on Earth as pulses of radio noise. Such objects are known as pulsars (fig. 91). Pulsars are often found at the sites of old supernova explosions, surrounded by expanding clouds of glowing material known as supernova remnants (fig. 90) – the debris from the explosion in which the pulsar was born. All pulsars are neutron stars, but there is no reason why all neutron stars should be pulsars.

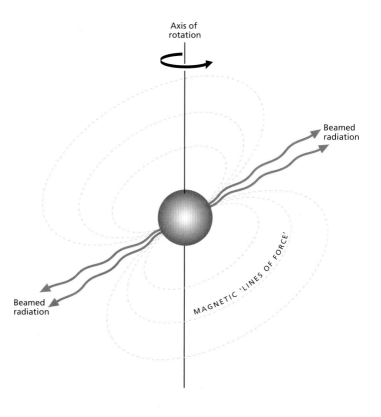

Fig. 91. *Schematic representation of a pulsar. Radio waves beamed out from the poles of a fast-spinning neutron star flick around the sky like the beams of a lighthouse.*

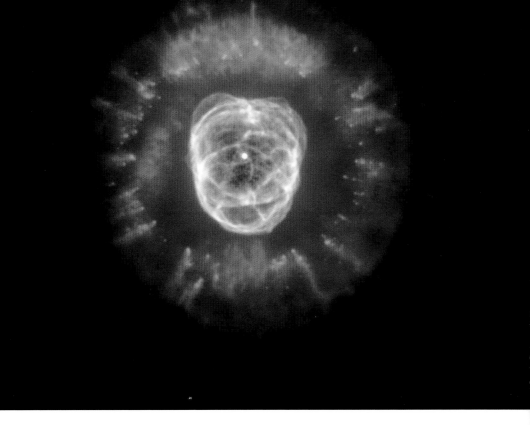

Black Holes and Supernovae

For a star that starts out with four to six times the mass of the Sun, the energy released in the final collapse blows away the outer layers of material and leaves the neutron star behind. For a star with six to eight times as much mass as our Sun, the final collapse releases so much energy that it overcomes the gravitational pull and the entire star is disrupted and blown away into space, leaving nothing behind. These extreme stellar explosions are known as supernovae. When a supernova explodes it can briefly shine as brightly as a whole galaxy of hundreds of billions of main sequence stars.

Some supernovae occur because a white dwarf star is orbiting a binary companion. In such a system, the gravity of the white dwarf can tug material off its companion star when the companion swells up to become a red giant. This stellar cannibalism means that the white dwarf steadily

Fig. 92. The 'Eskimo' Nebula, NGC 2392, as viewed by the Hubble Space Telescope in January 2000. The 10,000-year-old nebula is so nicknamed because, when viewed through ground-based telescope, it resembles a face surrounded by a fur parka. The 'parka' is really a disk of material embellished with a ring of comet-shaped objects, with their tails streaming away from the central, dying star. The nebula is about 5000 light years from Earth in the constellation Gemini.

gets more massive. If it starts out with a bit less than 1.4 solar masses of material, it will explode when its mass just reaches the Chandrasekhar limit. Because all these supernovae explode with exactly this critical mass, they all have the same brightness. They are known as Type 1a supernovae, and will come into the story of the Universe at large in later chapters.

Even a neutron star is not quite the end of the story of stellar collapse. If a stellar remnant is left with more than about three times as much mass as there is in the Sun today, with no possibility of generating heat in its core to hold itself up against the inward pull of gravity, then nothing can stop it collapsing into a black hole. The critical mass in this case is known as the Oppenheimer–Volkoff limit.[12]

Fig. 93. Artist's impression of a black hole in orbit around a supergiant star and 'feeding' off its companion.

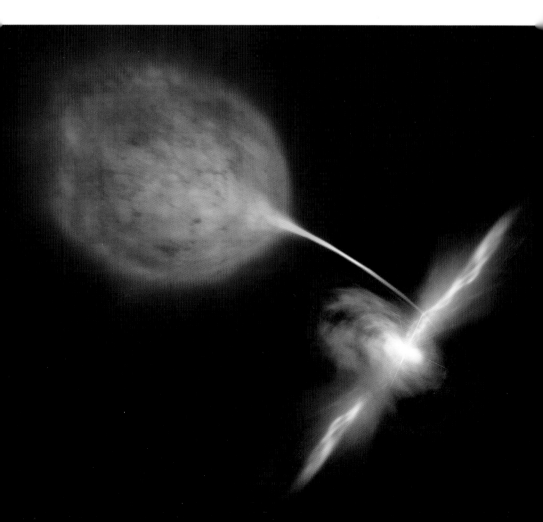

A black hole gets its name because nothing can escape from it, not even light, so we have no observations of what goes on inside a black hole, and have to rely on calculations based on Einstein's general theory of relativity. This tells us that the matter inside a black hole collapses all the way into a point with zero radius and zero volume, called a singularity. But even though we cannot see inside black holes, we can detect them because some of them are in binary systems, where the strong gravitational pull of the black hole is stripping matter from its companion star and swallowing it. The matter falling into the black hole gets hot and emits x-rays (fig. 93). Astronomers have found many x-ray sources of this kind, which are too small to be stars but that have several times the Oppenheimer–Volkoff limit of matter. They must be black holes.

There are over 500 active pulsars in our Milky Way Galaxy today, but such energetic emitters of energy can only be active for a relatively short time before they slow down and the radio noise from them fades away. The best estimate is that there are around 400 million 'dead' pulsars in our Galaxy. From calculations of how stars evolve and observations of the numbers of stars with different masses, astronomers know that there must be at least a quarter as many black holes formed by supernova explosions as there are neutron stars, so a conservative estimate is that there are 100 million black holes in our Galaxy. If they are scattered about randomly across the Galaxy, the nearest one is probably about 15 light years away, less than four times farther than the distance to the nearest star.

Many of these black holes will have been produced in the death throes of stars that start out with even more than eight solar masses of material. These are known as Type 2 supernovae, and they do not involve a slow build-up of mass towards the Oppenheimer–Volkoff limit, but the spectacular collapse of much larger amounts of material. They are the classic examples of stars that live fast and die young, not having time to move far from their birthplaces before they explode. Such stars are relatively rare, but they are interesting in their own right as well as being hugely important contributors to the processes that made the Solar System and ourselves what we are today (as we will explain in later chapters). So it is worth going into a little detail about how they die.

Seeding the Universe

By the time such a massive star has reached the stage of its life where an iron core is building up in its heart, there are many layers wrapped around the core. Just above the iron core itself, silicon is being converted into iron. In the next layer, oxygen and neon are being converted into silicon. Above that, carbon is being converted into oxygen. In the next layer, helium nuclei are fusing to make carbon. And in the outermost layer of the star there is still some hydrogen being turned into helium (fig. 94). Each layer also contains other elements, by-products of the nuclear fusion processes, such as sulphur, argon, nitrogen, chlorine, magnesium and many others familiar here on Earth. The process stops with the iron group of elements. Although energy is released when lighter nuclei fuse to make all the elements up to iron, in order to make these heavy nuclei fuse to make even heavier elements, energy has to be put in to squeeze them together. But a big squeeze is just what the material in the giant star is about to get.

Fig. 94. The 'onion skin' structure of a massive star late in its life.

Fig. 95 (facing) Supernova remnant LMC N 63A: the glowing shell created by the destruction of a massive star. The blue glow is from material heated to about 10 million degrees Celsius by a shock wave generated by the supernova explosion. The remnant is estimated to be in the range of 2,000 to 5,000 years old.

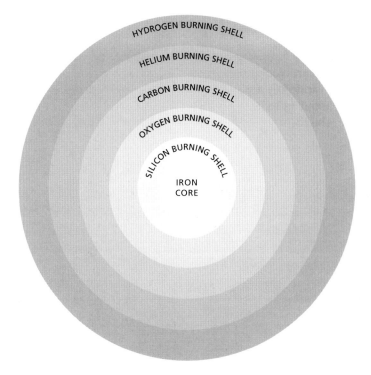

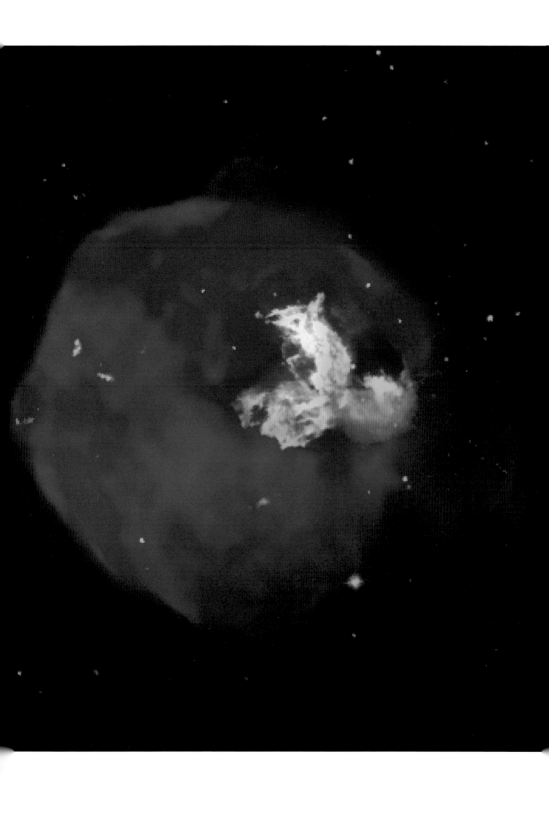

If the star has between about 15 and 20 times the mass of our Sun, the iron core itself must contain about as much mass as the Sun in a region the size of the Earth – rather like a white dwarf buried inside a star that extends overall about 50 times bigger than the Sun is today. By this stage of its life the star will have lost two or three solar masses of material, blown away into space. But this phase of its life lasts for but an eyeblink compared with the kind of timescales that apply to the evolution of the Sun. If a star starts out with seventeen or eighteen times as much mass as the Sun, it spends just a few million years on the main sequence and just a tenth as long as a helium-burning red giant. Carbon burning keeps the star going for only 12,000 years, the energy from fusion involving oxygen and neon holds it up for ten years, and the silicon lasts for only a few days. Then, nuclear fusion in the heart of the star ends and there is no longer a source of heat to provide the immense pressure required to hold the star up under its own weight. The core simply collapses, all the way down to a neutron star in a few seconds, releasing a huge burst of gravitational energy as it does so. For even larger superstars, with cores containing more than three times as much mass as the Sun, the collapse can go all the way into a black hole, releasing even more energy.

The energy emerges in the form of x-rays, gamma rays and fast-moving particles, travelling at a sizeable fraction of the speed of light, produced by the conversion of energy into mass. The total energy released in a few seconds is about a hundred times more than the total amount of energy radiated by a star like the Sun during its entire lifetime. The dozen or so solar masses of material spread over 50 solar radii in all directions above what has become an Earth-sized hole scarcely has time to start falling inwards before it is hit by a blast of energy and energetic particles from the collapsing core, producing a shock wave that ripples through the outer layers and pushes all the material, at least half the original mass of the star, away into space. The shock wave moves at about 2 per cent of the speed of light, and is so dense that it even absorbs neutrinos from the collapsing core, boosting its outward movement. Within the shock the extreme conditions trigger a profusion of nuclear interactions in which all the elements heavier than iron – things like copper, uranium, silver, mercury and lead – are manufactured (fig. 95).

This is where the chemical elements come from – they are manufactured inside stars and distributed through space in stellar winds and stellar explosions. Hydrogen and helium, themselves left over from the birth of the Universe, together make up 99 per cent of all the mass in the Universe that exists in the form of atoms and nuclei. Everything from lithium, the third-lightest element, to the iron group makes up 99.9 per cent of the remaining 1 per cent. But the total mass of everything heavier than iron adds up to just 0.1 per cent of the mass of the elements from lithium to iron – only 0.0001 per cent of the total mass of all the atoms and nuclei in the Universe. And everything heavier than hydrogen and helium has been made inside stars.

That really is the end of the story of stars; but it also takes us back to the beginning of the story, since new stars form out of the debris of these stellar explosions. It is also the beginning of the story of how stars interact with one another, forming islands in space called galaxies.

Fig. 96. Five young stars, seen in a small portion of the Orion Nebula viewed by the Hubble Space Telescope. Four of the stars are surrounded by gas and dust trapped as the stars formed, but left in orbit about the star. These are possibly protoplanetary disks, or 'proplyds', that might evolve into planets.

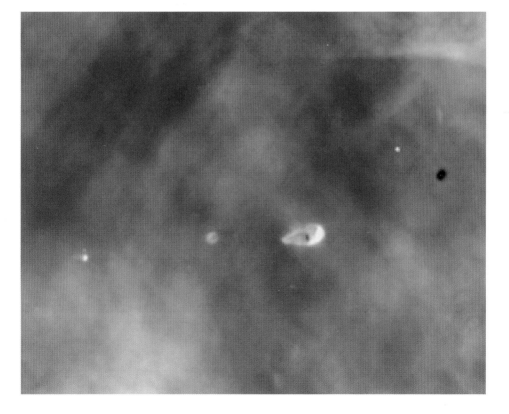

Chapter_Seven
The Milky Way and Other Galaxies

All of the stars we can see in the sky are part of an 'island' in space called a galaxy. Our particular island is known as the Milky Way Galaxy, or just the Galaxy, with a capital 'G'. It gets the name from a band of white across the night sky (in dark places far away from the glare of artificial lighting) stretching from horizon to horizon. To the ancients, this was an obvious and important feature of the night sky, which appeared to them like a misty or milky road or river among the stars. It was only with the invention of the telescope that astronomers, starting with Galileo, learned the true nature of the Milky Way – it is made up of enormous numbers of stars, so far away that they are too faint to be distinguished with the unaided human eye. He wrote in his book *The Starry Messenger* that the Milky Way is:

Fig. 97 (overleaf)
This infrared image shows hundreds of thousands of stars crowded into the Milky Way's swirling core. The region pictured here spans 890 light years sideways and 640 light years vertically.

Fig. 98. Artist's impression of our Milky Way Galaxy seen from above, with the approximate position of the Solar System indicated.

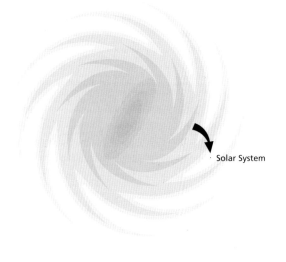

Solar System

In fact nothing but congeries of innumerable stars grouped together in clusters. Upon whatever part of it the telescope is directed, a vast crowd of stars is immediately presented to view. Many of them are rather large and quite bright, while the number of the smaller ones is quite beyond comprehension.

The stars of the Milky Way are no longer 'innumerable', but their numbers are still almost incomprehensible. By counting the stars visible in a small patch of the sky and multiplying up to take account of the size of the whole Galaxy, and other techniques based on measuring the mass of the Galaxy, astronomers estimate that our Milky Way Galaxy contains several hundred million stars (*fig. 97*). The Sun is an ordinary, unspectacular member of this array of stars, and it occupies no special place in the Galaxy, neither at the centre of the Milky Way nor at the edge (*fig. 99*).

Fig. 99. Artist's impression of our Milky Way Galaxy seen edge on, with the approximate position of the Solar System indicated.

Solar System

The Milky Way

Most of the stars in our Galaxy lie in a flat disc, about a hundred thousand light years across and 250 light years thick. The Sun is about two-thirds of the way out from the centre of the Milky Way, in the middle of the disc, so in whichever direction we look around the galactic equator we see the stars of the Milky Way. The band of stars is thicker and brighter towards the centre of the Galaxy, and thinner towards the rim, but we are completely surrounded by the stars of the disc. There is also a bulge of stars about 300 light years thick around the centre of the Galaxy, so that if we could see the Galaxy edge-on from outside it would look rather like two huge fried eggs stuck back to back (*fig. 99*). If we could view it from above, we would see that the brightest stars are not spread evenly across the disc, but form spiral arms trailing outward from the centre (*fig. 98*). Because of this, the Milky Way is classified either as a disc

galaxy or as a spiral galaxy; many other disc galaxies are known, both with and without spiral arms.

Around the Galaxy, spread over a spherical volume centred on the bulge of the Milky Way, there are many globular clusters of stars. Each globular cluster is a densely packed ball that may contain millions of individual stars, glowing like a jewel in the field of view of a large telescope (*fig. 100*). They are spread through a spherical halo around the disc of the Milky Way, and are not confined to the disc, but most of these clusters are no farther away from the centre of the Milky Way than we are. About 150 globular clusters have been identified and studied. They contain some of the oldest stars in the Galaxy, and at the heart of a globular cluster these stars are packed so densely together that there may be as many as a thousand stars within a volume only two or three light years across, smaller than the distance from the Sun to the nearest star to the Solar System, Alpha Centauri. The view of the night sky from a planet orbiting such a star would be spectacular; unfortunately, though, it is very unlikely that there are any planets orbiting these stars.

The stars in the central region of the Galaxy are also very old, and like the stars in globular clusters are known as Population II stars. These stars contain very little in the way of heavy elements, because they formed when the Universe was young, before the manufacture of heavy elements inside stars, stellar nucleosynthesis, got to work in a big way. Some of the stars in the disc are also old, Population II stars, but many are young, Population I stars, like the Sun. These stars contain much more in the way of heavy elements, because they formed relatively recently out of material enriched by previous generations of stars. Only Population I stars have planets associated with them, because planets are made of heavy elements.

At the very heart of the Milky Way, there is a black hole containing as much matter as 3 million stars like the Sun, or a whole globular cluster, but occupying a volume no bigger across than the distance from the Earth to the Sun. The black hole is revealed from studies of stars in orbit around it. Twenty or so stars near the galactic centre have been studied in detail, and found to be moving at speeds of nearly 10,000 km (6200 miles) per second in the grip of the intense gravitational pull of the black hole. Even though they are about 30 thousand light years away, these stars can

Fig. 100. A globular star cluster, called NGC 362, found in our own Galaxy. In this image, what look like foreground stars are actually part of a more distant neighbouring galaxy, the Small Magellanic Cloud.

be seen to have moved in photographs taken a few months apart over an interval of a few years; the closest of these stars to the centre approaches within 60 astronomical units of the black hole.

All of the stars in the disc of the Milky Way are also orbiting around the centre of the Galaxy, but at a much more leisurely rate. The Solar System, for example, is moving at roughly 250 km (155 miles) per second, and takes about 225 million years to travel once around the Galaxy. By studying the way stars at different distances from the centre of the Galaxy move, astronomers can work out how much matter there is altogether in the Galaxy holding the stars in its grip. This turns out to be about 1000 billion times the mass of the Sun, which is several times more matter than can be accounted for by bright stars, dust and gas in the Galaxy. This dark matter is spread more or less evenly through a spherical halo, extending far beyond the bright disc of the Milky Way, within which the Galaxy rotates. This is not unlike the way a thin layer of cream swirls around when it is stirred into a cup of black coffee.

Nobody yet knows exactly what the dark matter is. It does not interact with the kind of matter that stars, planets and people are made of except through gravity, and it cannot be any form of everyday matter, such as clouds of gas or lumps of rock. One of the most pressing problems of twenty-first-century physics is to identify the dark matter; all we can say now is that it exists, and that it is far more important in terms of its gravitational influence than the bright stuff.

Stellar Nurseries

It is the way the material in the disc of the Milky Way rotates in the grip of the dark matter that explains how new stars are born. There is a great deal more to the Milky Way than stars alone. The space between the stars isn't really empty. On average, there is just one hydrogen atom in every cubic centimetre of space between the stars of the Milky Way; but most of the interstellar material is concentrated in great clouds of gas that also contain tiny traces of dust, particles about the same size as the fine particles in cigarette smoke. In some places, the clouds of gas and dust are cold and dark, so they block out the light from stars behind them, and they look like black pits against the back-

Fig. 101. *A dramatic image of the Orion Nebula.*

ground of bright stars. In other places, the clouds are kept hot by the energy of stars shining inside them, and form colourful images in astronomical photographs; one famous example is the nearby Orion Nebula (*fig. 101*).

Even one hydrogen atom in every cubic centimetre adds up over a volume as big as the Milky Way, and it is estimated that there is about 3 billion solar masses of material in the form of molecular hydrogen and traces of other substances orbiting around the Galaxy in the region of the disc between the orbit of the Sun and the centre of the Milky Way. This is about 15 per cent of the mass of all the stars in the same region of the disc (excluding the central bulge) put together. In the large clouds, prosaically known as giant molecular clouds, there are between 1 billion and 10 billion atoms or molecules in every cubic metre, corresponding to thousands of atoms in every cubic centimetre. This is still almost empty compared with the air that we breathe, which contains tens of billion of billions of molecules in every cubic centimetre. But it is enough to make stars. The clouds contain material that has been thrown out by stars at the end of their lives, and material from the clouds is constantly being recycled into new stars, in a steady process that has gone on for billions of years.

We have already seen how a collapsing cloud of gas and dust can form stars (see Chapter Six, page 144). But what makes such a tenuous cloud collapse in the first place? The key to this continuous process of star formation and recycling lies in the spiral pattern of the Galaxy, and computer simulations show that the spiral pattern only persists because the whole disc is held in the gravitational grip of the large halo of dark matter. Eventually, the spiral pattern in a disc galaxy changes as a bar of stars grows across the middle of the galaxy; this is already happening in the Milky Way. Meanwhile, the spiral pattern that is characteristic of a galaxy like our own shows up so clearly because there are many hot, bright stars along the inner edge of the spiral arms. These stars are hot and bright because they are big, and, as we have seen, big stars burn out quickly; so they must also be young. They were born very close to where we see them now. They were born there because the spiral arm is actually a shock wave moving around the Galaxy, like a spiralling sonic boom. The shock wave is the important feature of the spiral pattern, but it moves more slowly than the clouds of gas and dust (and stars) that overtake it from behind. The shock wave moves at a speed of only about 30 km (18.5 miles) per second, while the stars and gas clouds move at speeds of 200 to 300 km (120 to 190 miles) per second, so they pile into the rear of the spiral arm and get squeezed by the shock wave as they pass through it. The best analogy is with the moving traffic jam that builds up behind a large, slow-moving vehicle on a motorway. The cars coming up behind the slow vehicle get squeezed into the outer lanes of the motorway and slow down as they go past, then continue on their way. At any minute, there is a traffic jam behind the heavy vehicle, but the cars in the traffic jam change as time passes.

Fig. 102. The Horsehead Nebula, also known as Barnard 33, is a cold, dark cloud of gas and dust, silhouetted against the bright background. The bright area at the top left edge is a young star still embedded in its nursery of gas and dust. The top of the nebula is also being sculpted by radiation from a massive star located out of the field of view.

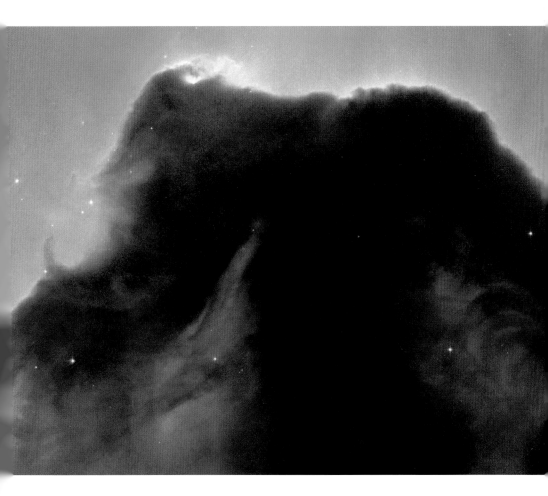

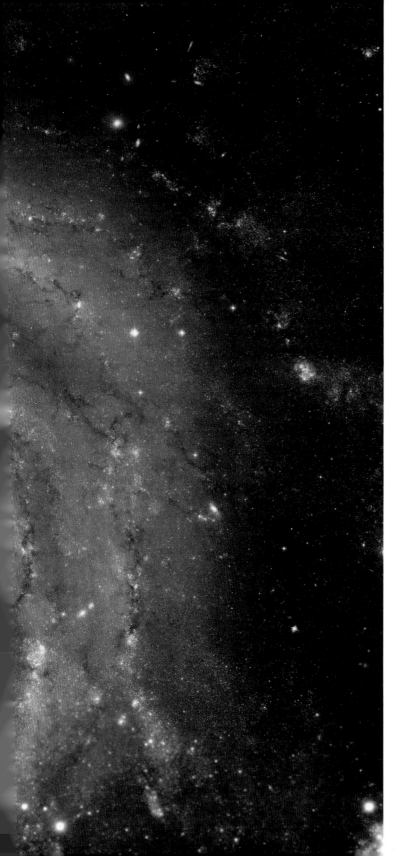

Fig. 103. *M101 galaxy, one of the best-known examples of 'grand design spirals', similar to our own Milky Way.*

It is the squeeze caused by the pile up in the traffic jam on the inner edge of a spiral arm that causes some of the molecular clouds to collapse and begin the process of star formation that we described earlier. The biggest stars formed in this way run through their life cycles very quickly and explode, scattering material back into interstellar space and making blast waves that help to maintain the spiral pattern. Many small stars, like the Sun, are made in this way and in the same places, but along with the surviving giant molecular clouds they go on quietly in their orbits for billions of years, passing through the spiral arms time and again on their travels.

But star formation is a rare process. Most clouds survive most encounters with a spiral arm. Only a few solar masses of material are turned into new stars each year by this process, while roughly the same amount of stuff is recycled back into space as old stars eject material. The biggest stellar explosions are the supernovae, and only two or three of these occur each century in a galaxy like the Milky Way. But since the Sun was born out of a cloud of interstellar

Fig. 104. Cassiopeia A – an artist's animation of a supernova explosion.

material 4.5 billion years ago, there have been several hundred million supernova explosions seeding the Galaxy with the raw materials for the formation of new planetary systems (*fig. 104*).

The Milky Way itself, though, is only about 10 to 12 billion years old, not much more than twice as old as the Sun, and even in 10 billion years making a couple of new stars every year only gives you 20 billion stars, even if you ignore the fact that a lot of them will have died by now. Where did all the rest of the stars in the Milky Way, at least ten times more than the number that could have been made in the time available by this steady process, come from? The answer is that they must have formed in a burst of activity (perhaps more than one burst) when the Galaxy was young; we know this because we can see such starbursts going on in other galaxies. Studying other galaxies also gives us a clear idea of the place of the Milky Way in the Universe. Just as the Sun is an ordinary star, it turns out that the Milky Way is just an ordinary galaxy.

An Average Galaxy

Until the 1920s, people thought that what we now call the Milky Way Galaxy was the entire Universe. At that time, accurate distances to even the nearest stars had only been known for about 90 years, and astronomers were still struggling to come to terms with the size of the Milky Way itself. In 1920, the American astronomer Harlow Shapley (1885–1972) used measurements based on studies of Cepheids and globular clusters to work out that the Milky Way must be about 300,000 light years across, and that the Sun is located far out from the centre. He thought that this was the entire Universe, but other astronomers were puzzling over the nature of fuzzy patches of light, called nebulae, that they could see through their telescopes and in astronomical photographs.

Some nebulae turned out to be clouds of gas and dust within the Milky Way, often the remnants of dying stars. But some were not. Could they be other galaxies, islands in space beyond the borders of the Milky Way? The puzzle was partly solved a few years later, when another American astronomer, Edwin Hubble (1889–1953), identified Cepheids

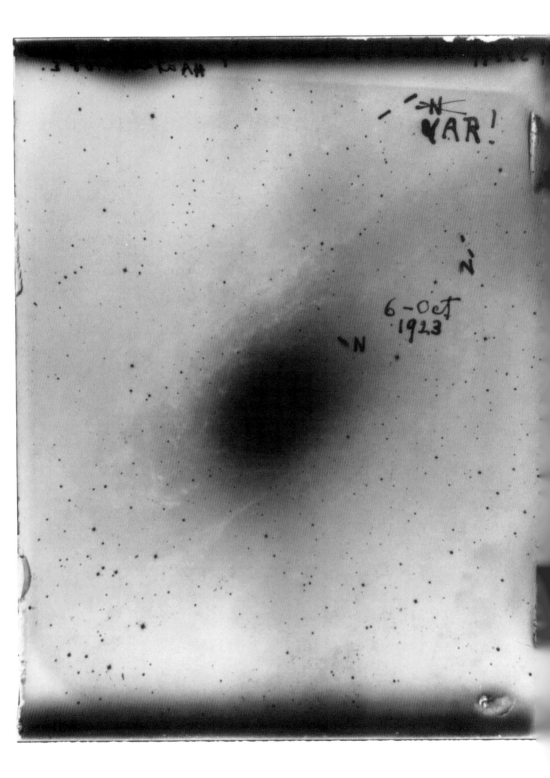

in what was then called the Andromeda Nebula (*fig. 106*), but is now known as the Andromeda Galaxy. Hubble's measurements implied that this 'nebula' was a star system a million light years away, a distance three times greater than the size of Shapley's 'universe'. All of these numbers have since been revised, because there was an error in the early calibration of the Cepheid distance scale. We now know that the Milky Way is about a third of the size Shapley calculated, and that the Andromeda Galaxy is a bit more than twice as far from us as Hubble thought. But what matters is that by the end of the 1920s astronomers knew that there were other spiral galaxies extending far out into the Universe beyond the borders of the Milky Way.

At first, because of the problems with calibrating the distance scale, it still seemed that there was something special about the Milky Way. Our home Galaxy seemed to be far bigger than any of the others, and was sometimes described as like a continent surrounded by islands. The only way to estimate the sizes of galaxies is to measure their distances from us, using Cepheids or techniques based on how bright (or faint) supernovae and other bright objects seen in them look to us. Then you can compare these distances with the angular size of the galaxy on the sky. A galaxy may look small because it really is small, and is close to us, or because it is big and far away. The farther away a galaxy is, the smaller it looks.

Right up until the end of the twentieth century, many astronomers thought that the Milky Way was a larger than average disc galaxy. But measurements of Cepheids in other similar galaxies made using the Hubble Space Telescope gave very accurate distances to all the nearby spirals that resemble the Milky Way. That made it possible to work out all their sizes and compare these with the size of the Milky Way. It turns out that our Galaxy is almost exactly the average size of all such spiral galaxies, and is actually smaller than the Andromeda Galaxy.

Islands in Space

But there are other kinds of galaxy in the Universe. The most important distinction is between disc galaxies (*figs. 103, 107*) and so-called elliptical galaxies (*fig. 108*). An elliptical galaxy looks like a circular or elliptical patch of light

Fig. 105 (overleaf)
The Hubble Space Telescope has been used to monitor Cepheid variables in the barred spiral galaxy NGC 1672. Clusters of hot young blue stars are clearly visible along its spiral arms.

Fig. 106 (facing)
A Cepheid variable appears alongside novae in the Andromeda Nebula M31, captured by Edwin Hubble in this 1923 photograph, taken with the 100- inch telescope. He marked the Cepheid as 'Var'.

in astronomical photographs, with no sign of a surrounding disc. Their shape is like a squashed sphere, and they are made up almost entirely of old, red stars. Unlike spiral galaxies, they are not making new stars today. The central bulge of stars seen in a disc galaxy like the Milky Way looks very much like an elliptical galaxy; all the evidence suggests that such bulges formed in the same way as ellipticals, growing around a central black hole, but that not all of the young galaxies developed surrounding discs.

Elliptical galaxies come in a wide range of sizes. The smallest ones have about a million times as much mass as our Sun, and are rather like globular clusters. The largest ones are the biggest galaxies known, and have as much mass as several thousand billion times the mass of our Sun, roughly ten times more than the mass of an average disc galaxy like the Milky Way. These very large galaxies have

Fig. 107. *A view of the disk galaxy NGC 5866, which happens to be tilted nearly edge-on to our line-of-sight. Hubble's image reveals a crisp dust lane dividing the galaxy into two halves and showing its structure: a subtle, reddish bulge surrounding a bright nucleus, a blue disk of stars running parallel to the dust lane, and a transparent outer halo. Our Galaxy would look like this if viewed edge-on from the outside.*

formed from repeated mergers, when two galaxies collide and their stars get mixed up into one system. Even when disc galaxies collide with one another, the end product will be an elliptical galaxy, because all of the structure in the discs is destroyed. Individual stars do not collide with one another in galactic mergers, because the distances between stars are so much greater than the sizes of stars themselves. Instead, gravitational forces change the orbits of the stars and mix them up. In some large elliptical galaxies, streams

Fig. 108. Complex loops and blobs of cosmic dust lie hidden in the giant elliptical galaxy NGC 1316. This image made from data obtained by the Hubble Space Telescope suggests that it was formed from a past merger of two gas-rich galaxies.

of stars left over from the merging process can still be identified, fossil remnants of discs that used to be. In other cases, galaxies are caught in the act of merging, and very often huge bursts of star formation are seen where these mergers have sent shock waves rippling through the clouds of gas and dust between the stars. These are known as starburst galaxies (*fig. 109*).

Our Galaxy and the Andromeda Galaxy are moving towards one another at hundreds of kilometres per second, measured by the Doppler effect. They will collide and merge, undergoing a burst of star formation in the process, within the next 10 billion years. There will be two less disc galaxies in the Universe after the merger, and one more large elliptical.

Fig. 109. The starburst galaxy Messier 82 (M82) is remarkable for its webs of shredded clouds and flame-like plumes of glowing hydrogen blasting out from its central regions where young stars are being born 10 times faster than they are inside in our Milky Way Galaxy.

Any galaxy that cannot be classified as a disc galaxy or an elliptical is simply called an irregular galaxy. Irregulars make up about 10 per cent of all known galaxies, while 60 per cent are ellipticals and the rest spirals. The visible Universe is estimated to contain several hundred billion galaxies, very roughly the same as the number of stars in the Milky Way. Our place in the Universe is truly insignificant – except to us.

Across the Universe

These galaxies are only visible at all because the distances between galaxies, in proportion to their size, are much smaller than the distances between stars, in proportion to their size. Apart from systems where two or more stars orbit around one another, the distances between stars are typically tens of millions of times bigger than the diameters of the stars themselves. A useful way to picture this is to imagine a star like the Sun shrunk down to the size of an aspirin. On this scale, the nearest star to the Sun would be represented by another aspirin 140 km (87 miles) away. No wonder stars do not collide with one another even when galaxies merge! Changing the analogy slightly, the number of stars in the Milky Way is about the same as the number of rice grains that could be packed into St Paul's Cathedral in London, if the rice grains were touching their neighbours. On this scale, if you spread the grains out to make a proper scale model of the Galaxy, it would stretch from here to the Moon, a distance of 400,000 km (250,000 miles).

But if we represented the whole Milky Way Galaxy using a single aspirin, the Andromeda Galaxy, our nearest comparable neighbour, would be represented by another aspirin just 13 centimetres (5.1 inches) away. The nearest large grouping of galaxies, called the Virgo Cluster, would be represented by between two thousand and three thousand aspirin spread over a volume only as big as a basketball, at a distance of three metres (9.8 feet). And the entire observable Universe, with all its hundreds of billions of galaxies, would be only a kilometre or so across. Turning this around, if the spaces between galaxies were in the same proportion to their diameters as the spaces between stars are to stellar diameters, the distance to the nearest large galaxy would be at least 10 million times the diameter of the Milky Way, which would place the Andromeda Galaxy 1000 billion light years away. It would be beyond the range of any telescope, and it would look to us as if the Milky Way really was the only large galaxy in the Universe.

But because the Universe is, as far as galaxies are concerned, a crowded place, there's a lot going on on scales larger than galaxies that we can still study in detail using telescopes on Earth and instruments carried aloft on satellites. We have already mentioned the most important feature of this view of the Universe beyond the Milky Way – galaxies do not occur in isolation, but form groups known as clusters, like the Virgo Cluster (*fig. 111*). Our own Galaxy is a member of a small group of galaxies, known simply as the Local Group, which has about forty members, bound to each other by gravity (*fig. 110*). These include the Andromeda Galaxy, but most members of the Local Group are small irregular galaxies. The Virgo Cluster is the nearest large cluster of galaxies to us, but still at a distance of more than 50 million light years. It contains at least 2500 galaxies, and unusually most of these (perhaps three-quarters) seem to be disc galaxies.

Fig. 110. *The Local Group of galaxies. The Milky Way is not actually at the centre of the Local Group, but this diagram has been drawn from our own perspective, just as the first Western maps of the globe placed Europe in the middle.*

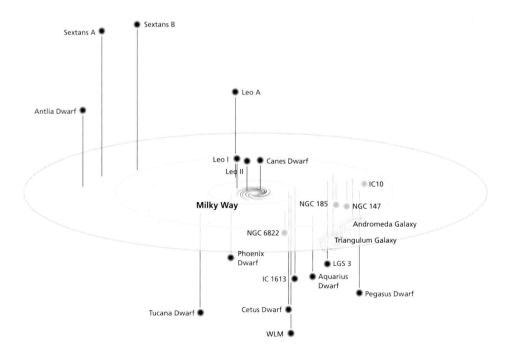

Clusters of galaxies also group together in larger units, called superclusters. The Local Group is part of the same supercluster as the Virgo Cluster, and this is sometimes referred to as the Local Supercluster. The hierarchy continues upwards. Superclusters themselves are linked in chains and filaments that stretch across the observable Universe. Photographic maps of the positions of millions of galaxies on the sky show a foamy pattern of bright filamentary structures surrounding darker voids where there are few galaxies. Statistically speaking, though, the pattern is the same in all directions, so on average the Universe would look the same from any galaxy; there is no special place in the Universe.

Fig. 111. Ultraviolet image *of a small portion of the* Virgo Cluster of galaxies.

Galaxies with Attitude

But there are unusual galaxies. Apart from the starburst activity associated with mergers, some galaxies have active hearts that are the source of great outbursts of energy. It used to be thought that there were many different kinds of active galaxy, but it is now understood that all of this activity is associated with black holes, and the difference is only a matter of degree.

The black hole at the centre of the Milky Way contains only about 3 million solar masses of material, on the small side as such things go, and it is not very active today, unlike the black holes in the nuclei of most galaxies. But the black

Fig. 112. A region of star birth known as NGC 604, in the galaxy M33, one of the nearest galaxies to the Milky Way. This is similar to star-birth regions in the Milky Way, such as the Orion Nebula, but it is much larger and contains many more recently formed stars.

holes in the nuclei of some other galaxies are estimated to have masses of hundreds of millions of times the mass of our Sun, still contained within a volume no bigger across than our Solar System. Even such large black holes – large for a black hole, but still only about a thousandth of the mass of a galaxy like the Milky Way – will be fairly quiet if they do not have a source of matter to feed on, but if material falls into the black hole it can produce outbursts of various degrees, depending on how much matter it swallows. These active galaxies are grouped together under the name Active Galactic Nuclei, or AGN, which includes objects known for historical reasons as quasars, radio galaxies, Seyfert galaxies, N galaxies and BL Lac objects. The energy that powers AGN comes from the gravitational energy released when matter funnels into the hole, and that energy escapes because black holes are messy eaters.

The infalling matter speeds up as gravitational energy is turned into kinetic energy, just as a falling object speeds up if you drop it from a cliff. But it doesn't fall straight into the hole. The key to the activity of AGN is that although the gravitational pull of the black hole is very strong, the surface of the hole is very small, so everything piles up at the entrance before it is swallowed. Because of rotation, it forms a swirling disc orbiting around the hole while material funnels in to the hole itself from the inner edge of the disc. The fast-moving particles of matter collide with one another in the disc, generating enormous heat, which can show up as x-rays, radio waves or visible light. Matter from the disc falls in toward the hole along the equator, but the energy released near the hole can send blasts of material shooting out from the poles, at right angles to the equator. This sometimes shows up as bright jets of matter moving outward at a sizeable fraction of the speed of light. When electrons and other charged particles ejected in this way interact with magnetic fields around the galaxy, they can produce two bubbles of radio noise, one on either side of the galaxy.

If all of a lump of matter with mass m could be converted into pure energy, the energy released would be mc^2, in line with Einstein's famous equation, where c is the speed of light. When four hydrogen nuclei fuse to make one helium nucleus (the process that keeps the Sun shining), just 0.7 per cent of their mass is released as energy. The immense

13___Rest mass energy is the energy you would get if all the mass (*m*) of an object were turned into energy, in line with Einstein's famous equation $E=mc^2$.

gravitational field of a supermassive black hole is so powerful that the equivalent of as much as 10 per cent of this rest mass energy[13] of the infalling matter can be released. If a black hole had a mass of 100 million times the mass of the Sun – by no means the largest known – and it swallowed a couple of solar masses of matter each year, perhaps literally in the form of stars ripped apart by its gravitational field, it would release enough energy to explain the output of the most active AGN we have seen.

But this cannot go on forever. It is thought that all galaxies except the small irregulars go through a phase of activity like this when they are young, with the strength of their activity depending on the size of the black hole in the nucleus. When all the material near the hole has been swallowed, however, it settles down into a quiet old age, like the one at the heart of the Milky Way. This may not, though, be the end of the story. A close encounter with another galaxy, or a merger like the one that will eventually take place between our Galaxy and the Andromeda Galaxy, will stir up the material in the galaxy and spill a fresh supply of gas, dust and stars in towards the black hole.

This picture is confirmed because we know that there were more AGN, and they were more active, long ago when galaxies were younger. We know this because of one of the most useful features of telescopes – they act, in effect, as time machines.

Light travels at a finite speed, so it takes a finite time to cross space. We see the Sun by light that left it just over eight minutes ago, and we see a star that is, say, 10 light years away as it was 10 years ago. When we look at a galaxy that is a billion light years away, we see it as it was a billion years ago, before life on Earth had emerged from the ocean on to the land. For distant galaxies, these times are so big that there are complications in exactly how you work out the distances to galaxies across the Universe, because, as we discuss in the next chapter, the Universe is expanding. This means that the distance between two galaxies is not the same now as it was when light set out on its journey. So astronomers generally prefer to refer to the 'look back time' to distant galaxies, a terminology that emphasises the time travel aspect of their observations. The farthest galaxies we can see are at distances corresponding to look back times of around 13 billion years, where we see the brightest AGN

Fig. 113. *These 'fingers' are columns of cool interstellar hydrogen gas and dust that are also the places where new stars form. They are part of the Eagle Nebula, a star-forming region 7000 lightyears away in the constellation Serpens. The tallest column (left) is about a lightyear long from base to tip; a system like our Solar System forms from the tiniest tendril on the end of a column.*

(some of the kind known as quasars) and the largest proportion of AGN compared with quiet galaxies.

All of this shows that the Universe has changed as it has got older. Studies of galaxies combine with other evidence to tell us that there was a definite beginning, the Big Bang, in which the Universe as we know it was born, about 13.7 billion years ago. The study of the Big Bang and the subsequent evolution of the Universe – cosmology – is one of the most important areas of astronomical research today. It can also shed light on the origin of galaxies.

Chapter_Eight
Cosmology

Edwin Hubble (fig. 115) made the two most important discoveries in cosmology. First, as we have seen, he proved that many nebulae are other 'island universes' beyond the boundaries of the Milky Way. Then, he discovered that the galaxies are moving apart from one another – that the Universe is expanding. But he didn't make the second discovery on his own; the astronomer who actually carried out most of the painstaking observational work was Hubble's colleague at the Mount Wilson Observatory, Milton Humason (1891–1972). Hubble, the more senior astronomer, chose Humason as his partner because Humason was, quite simply, the best observer in the world in the 1920s, and able to push the abilities of what was then the best telescope on Earth, the 100-inch (2.54 m) reflector at Mount Wilson, to the limit.

From the perspective of the twenty-first century, it is worth taking stock of what those measurements involved. Humason had to spend long hours in the cold of an unheated telescope dome, open to the sky, on a mountain top at night, keeping the telescope trained on a particular patch of the sky while light from a distant galaxy was focused on to a glass photographic plate. The dome had to be unheated, because convection currents from any heater would disturb the air and distort the image; and observations were best made in winter, when the air was still and the nights were long. Often, the galaxy being studied would be so faint that at the end of one night's observing, the photographic plate would have to be packed away, in the dark, in a light-proof box, then taken out the next night and used to build up a brighter image of the distant galaxy. And then the plate had to be developed by hand before its image could be analysed. All this required enormous patience – a quality Humason had in abundance, unlike Hubble – as well as great skill. It is a far cry from modern techniques, where telescopes can be operated by remote control from air-conditioned rooms, and the light is gathered on photographic chips, with photons (particles of light) being counted by computers.

In the second half of the 1920s, Hubble was still primarily interested in measuring the distances to galaxies. He was intrigued by a discovery that had been made in the previous decade by Vesto Slipher (1875–1969), an astronomer working at the Lowell Observatory in Flagstaff, Arizona. Slipher had been working with a 24-inch (61 cm) refracting telescope that had a new instrument called a spectrograph attached to it. This could make photographs of the spectra of faint astronomical objects, by adding up the light over several nights if necessary. Among the objects Slipher studied in this way were several of the family then still known as nebulae, which Hubble was about to prove were actually

Fig. 115. Edwin Hubble.

external galaxies. By 1925, just when Hubble was beginning to measure distances to galaxies, Slipher had measured 41 of these spectra, and found that just two of them (including the Andromeda Nebula) showed blueshifts, while 39 showed redshifts. This was the limit of what he could do with the 24-inch telescope, but the evidence hinted that the galaxies that looked bigger and brighter had smaller redshifts. The obvious inference was that galaxies that look bigger and brighter are closer to us – so Hubble guessed that measuring redshifts might be a way of measuring distances to galaxies, and roped Humason in to test the idea with the 100-inch telescope. Humason measured the redshifts, while Hubble estimated the distances to the same galaxies, using the techniques we described in the previous chapter.

By the beginning of the 1930s, Hubble and Humason had made enough observations to show that the relationship between redshift and distance is about as straightforward as it could possibly be: the redshift is proportional to the distance – or, putting it the way round that mattered to Hubble, distance is proportional to redshift (*see fig. 116*). This is now known as Hubble's Law. It means that if one galaxy has twice the redshift of another it is twice as far away, and so on. Once the distances to a few nearby galaxies had been measured by other means, the rule could be

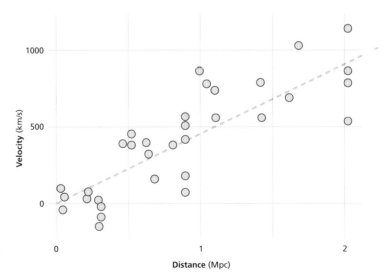

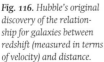

Fig. 116. *Hubble's original discovery of the relationship for galaxies between redshift (measured in terms of velocity) and distance.*

calibrated, and distances to other galaxies, much farther away across the Universe, could be measured simply by measuring their redshifts. In fact, this simple law only applies accurately to relatively nearby galaxies, and a more subtle relationship applies farther out across the Universe, but this does not detract from the importance of Hubble's discovery (fig. 117)

Hubble himself was not interested in why the light from galaxies showed redshifts. All he cared about was how the redshift (whatever its cause) could be used to measure distances. But the natural guess people made at first was that the redshifts are caused by the Doppler effect. If so, it meant that just two external galaxies (including Andromeda) are moving towards us, and all the rest are moving away – not as individuals, but as members of clusters like the Virgo Cluster. It was soon realised, however, that this recession of the galaxies is not caused by galaxies and clusters moving through space. Albert Einstein's (fig. 119)

Fig. 117. *The Sloane Sky Survey map of galaxies. Galaxies identified on the sky (right) have their distance determined from their redshift to make a map 2 billion light years deep (left) where each galaxy is shown as a single point, the colour representing the luminosity. In this map, the distance from the central point represents the distance to the galaxy from the Milky Way. This shows only those 66,976 out of a total of 205,443 galaxies surveyed that lie near the plane of Earth's equator.*

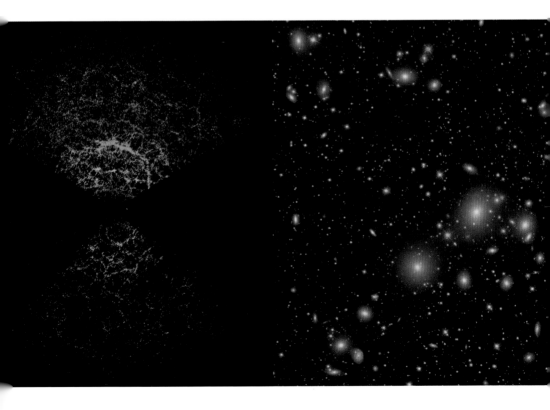

general theory of relativity, which he had completed in 1915, described how space itself could be bent by the presence of matter, like a stretched rubber sheet with a heavy weight on it. The equations also described how space as a whole could stretch, but in 1915 nebulae hadn't even been identified as external galaxies, and Einstein had dismissed this as a trick of the mathematics with no physical significance. After the discovery of the redshift–distance relationship, Einstein and other mathematicians realised that this was exactly what his equations described – space itself stretching and carrying clusters of galaxies along with it. This was the beginning of modern cosmology.

The cosmological redshift is not a Doppler effect. It is not caused by galaxies moving through space, but by the space between the galaxies stretching during the time it takes light to get from one galaxy to another, and stretching the light to longer wavelengths. Galaxies do move through space, producing Doppler effects in their spectra, but these are simply added to or subtracted from the cosmological redshift – which is why, for example, Andromeda shows a blueshift. It is moving towards us through space faster than the space between us and Andromeda is expanding. But for all except the nearest galaxies, the cosmological redshift dominates.

The usual way to get a picture of what is going on is to imagine a perfectly round rubber ball with spots of paint

Fig. 118 (left) Physicists John Cockcroft (left) and George Gamow, c. 1930.

Fig. 119 (right) Albert Einstein shortly after the publication of his General Theory of Relativity.

Fig. 120. *Robert Wilson and Arno Penzias, photographed in 1978, the year of the award of their Nobel prize for the discovery of the cosmic microwave background radiation, at the Horn Antenna in Holmdel, New Jersey.*

spattered on it. If the ball is pumped up so that it expands, all the spots will move apart from one another, not because they are moving through the rubber skin of the ball, but because the skin is stretching. And, crucially, there will be no centre to the expansion – from whichever spot you choose to measure, you will see all the other paint spots receding, exactly in line with Hubble's Law. Once again, we see that there is nothing special about the Milky Way. The view of the universal expansion would be the same from any galaxy in the Universe.

Fig. 121. *Like the expanding surface of a sphere, there is no centre to the expansion of the Universe. From any cluster of galaxies, the view is of all the other clusters receding.*

The startling implication of this discovery was that if the Universe is getting bigger now, it must have been smaller in the past. Go back far enough, and you would come to a time when all the galaxies were piled up on top of each other. If you imagine winding the expansion backwards even further, you would come to a time when the whole Universe was concentrated at a point. Using Hubble's Law and the latest calibration of the redshift–distance relation, we can calculate that this corresponds to a time about 14 billion years ago. This realisation is what led to the idea that the Universe emerged from a hot fireball, the Big Bang, and has been expanding and cooling ever since. But the Big Bang model of the Universe really only began to be taken seriously in the 1960s, when radio astronomers Arno Penzias and Robert Wilson (*fig. 120*) discovered a weak hiss of radio noise that they identified as the fading echo of the Big Bang itself.

This background radiation is in the microwave part of the spectrum, and comes from all directions in space (*fig. 122*), so it is known as the cosmic microwave background radiation, or just as the cosmic background radiation (CBR). Different parts of the electromagnetic spectrum correspond to radiation with different temperatures. If the radiation peaks in the part of the spectrum corresponding to the yellow-white of the Sun, this is equal to a few thousand degrees on the Kelvin scale; a red-hot lump of iron is cooler, the invisible infrared radiation you feel if you hold your hand near a radiator cooler still, and so on. Similarly, objects that emit mainly ultraviolet radiation or x-rays are

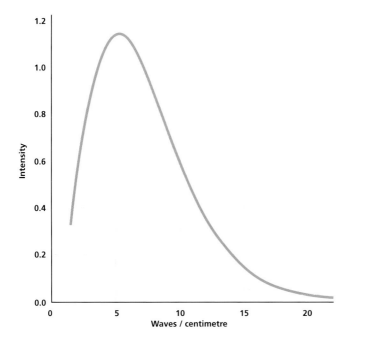

Fig. 122. *Cosmic microwave background spectrum. The spectrum follows a perfect 'black body' curve corresponding to radiation with a temperature of 2.725 K.*

much hotter than the Sun. The temperature of the cosmic background radiation, however, is just 2.7 K, or −270.5 °C. But it fills the entire Universe.

The explanation for the existence of the CBR is that the Universe began in a hot fireball, much hotter than a star. As it expanded, the fireball cooled, in much the same way that gas expanding out of an aerosol spray can cools. At first, radiation (photons) bounced around between charged particles in the same way that photons bounce around in the heart of the Sun. But when the whole Universe cooled to the point where charged particles got locked up in electrically neutral atoms, the radiation could range freely through space, just as it escapes from the surface of a star. Inevitably, this happened when the temperature of the entire Universe was the same as the temperature at the surface of a star today, a few thousand degrees. When the electromagnetic radiation escaped, a few hundred thousand years after the beginning of the expansion, it was very much like sunlight. But since then, it has been redshifted by the expansion of the Universe and stretched to longer wavelengths, turning it into microwaves. The entire Universe is now filled with this radiation, like a very cold microwave oven. When we use radio telescopes to 'look' at the cosmic background radiation, we are seeing a highly redshifted but direct view of the Big Bang itself, using 'light' that was radiated more than 13.5 billion years ago.

Of course, this understanding of the CBR did not emerge overnight. Following the discovery of the background radiation in the 1960s, it was studied in detail, first using radio telescopes on the ground and then using satellites designed specifically for the job. It is only in the past ten years or so that the more detailed studies of the background radiation made using satellites have shown that it is not *quite* perfectly uniform, and have used it to tell us more about the nature of the Universe we live in, and where structures like galaxies come from.

Just before the Universe cooled to the point where radiation stopped interacting with matter, the radiation and the matter were coupled together very strongly, like milk spread evenly through a cup of tea. Where the matter was a little less dense than the average, the radiation could cool off slightly, but where the matter was a bit more compressed than average it would be at a slightly hotter temperature. After the radiation and the matter decoupled, these differences in temperature remained imprinted in the radiation, while the matter got clumped together under the influence of gravity, collapsing into sheets and filaments within which clusters of galaxies formed. Stretching the analogy slightly, the picture is a bit like the way milk that is slightly 'off' forms little lumps when added to tea, instead of being spread out evenly. If the Universe had been perfectly smooth when the matter and radiation decoupled, then the CBR would be perfectly smooth today. But there would be nobody around to notice, since if this had been the case matter could never have clumped together to form galaxies and we would not be here.

The fluctuations required to do the job are so tiny, though, that for a long time it seemed unlikely that they would ever be measured. When satellites became sophisticated enough to do the job, in the first decade of the twenty-first century, they found that the average temperature of the CBR is 2.725 K, and that the fluctuations range from 2.7251 K to 2.7249 K. This may seem almost insignificant but, in fact, the detailed studies of the CBR show exactly the right pattern of hotter and colder patches to match the pattern of density variations that would lead to the kind of structure we see in the Universe around us.

But there is one more thing required to do the job of making galaxies. We have already mentioned that there must be a lot of dark matter in the Universe to explain the nature of galaxies like the Milky Way. Even more dark matter is needed to explain why clusters of galaxies don't fly apart. Clusters of galaxies are like swarms of bees, with each individual moving around within the cluster under the influence of gravity, while the whole cluster moves as a unit, carried by the universal expansion. By measuring the velocities of individual galaxies within clusters using the Doppler effect, it is straightforward to calculate how much mass there must be in the cluster to stop the galaxies escaping. This is always a lot more than the amount of matter we can actually see in bright galaxies in the cluster. The background radiation independently tells us the same thing. Computer simulations of how structures can develop in the Universe as it expands show that the pattern of galaxies and clusters we see can only have grown from fluctuations the size of the ones revealed by the background radiation if

14__Some of the atomic matter (also known as baryonic matter) is itself dark, like the dust between the stars. But our understanding of the physics of the Big Bang sets a limit on how much of this baryonic matter there can be. What astronomers call dark matter is different stuff entirely.

there is about six times as much dark matter as there is everyday atomic matter.[14] Then, everything fits together beautifully.

Even this, though, is not the end of the story of what we can learn from the CBR. In the 1930s, soon after the expansion of the Universe was discovered, cosmologists began to puzzle over the question of whether the expansion will continue forever, or whether it will stop one day, and perhaps even go into reverse. The answer depends on the way space is curved, as defined by the general theory of relativity.

There are three possibilities (*fig. 123*). If there is more than a certain amount of matter in the Universe, corresponding to a particular density at the present day, then three-dimensional space is curved in the same way that the two-dimensional surface of a sphere is curved, so that it folds back on itself and has no edge. In such a situation, if you

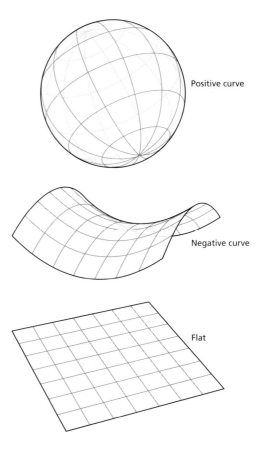

Positive curve

Negative curve

Flat

Fig. 123. *The possible shapes of the Universe, according to the general theory of relativity.*

head off in one direction and keep going long enough, you will end up back where you started, just as if you keep going in the same direction on the surface of the Earth you will get back to where you started after travelling right round the globe. This is called a closed universe, and if the real Universe is like that it will expand for a while, slow down as gravity works against the expansion, and then collapse back on itself. This is like the way a ball thrown straight up in the air slows down under the influence of gravity and falls back to Earth.

At the other extreme, if there is less than the critical amount of mass in the Universe, it is said to be open, and will expand forever. This is like the way in which a spacecraft launched with sufficient speed will escape entirely from the Earth's gravity. The geometry is harder to picture in this case but it is the three-dimensional equivalent of a saddle surface, or a mountain pass, extending out to infinity in all directions.

Just at the dividing line between these two possibilities is the third alternative, that space is flat in three dimensions in the same way that a piece of paper smoothed out on a table is flat in two dimensions. If the real Universe is like that, it will keep expanding but at a slower and slower rate until eventually it hovers on the brink of collapse, but never collapses. At least, it used to be thought that that was the fate of a flat universe. Now, cosmologists have another idea. If there was a curvature to space, it would bend light, like a lens. As a result, the images of very distant objects and the pattern of irregularities in the cosmic background radiation would be distorted in a particular way if the Universe is closed, and in a different way if the Universe is open. The observations do not reveal any trace of such distortions, so cosmologists are sure that the Universe is flat. This means that there must be a certain amount of mass in the Universe, which translates as a certain density today. But the amount of matter in the Universe, adding together baryonic matter and dark matter, provides only about 27 per cent of this critical density. So observations of the cosmic background radiation tell cosmologists that there must be another form of mass dominating the Universe. This is called dark energy. Just as all mass has an energy equivalent, so all energy has a mass equivalent, and although dark energy is not matter it has mass and affects the curvature of space and the way the Universe expands. The fact that space is flat tells us how much mass-energy there must be altogether; if 27 per cent is matter, then 73 per cent must be dark energy.

Dark energy shows its influence on the Universe directly by the way it affects the expansion. When the distances to remote galaxies are measured using observations of supernovae, it turns out that these galaxies are all a little bit farther away from us than they should be according to the simplest interpretation of their redshifts. Everything falls into place, however, if the expansion of the Universe is speeding up, so that distant galaxies are a bit farther away than the simple Hubble's Law implies. The effect is too small to be measured for nearby galaxies, which is why it was not noticed until the end of the twentieth century. It is thought that dark energy acts as a kind of antigravity, stretching space, and that this effect will get bigger as the Universe ages – as galaxies move farther apart, their gravitational bonds weaken, but the dark energy keeps on pushing. If these new ideas are correct, then space will always be flat, but the expansion of the Universe will get faster and faster until, in about 100 billion years from now, galaxies are so far apart that it will be impossible to see anything beyond the Milky Way and its companion galaxies in the Local Group. What happens then is a matter of guesswork.

How the Universe actually began is also a matter of guesswork; but it is astonishing how far back in time we can go before cosmologists have to resort to educated guesswork. Because we know the temperature of the background radiation today, and the overall density of the Universe, and how fast the Universe is expanding, we can calculate backwards in time to work out what the temperature and density were at any time in the past. We know that the Universe as we know it began just under 14 billion years ago in the Big Bang, and one of the objectives of cosmology is to work out what conditions were like as close as possible to that beginning, which is sometimes referred to as 'time zero'.

The most extreme conditions of density that exist on Earth today are in the nuclei at the hearts of atoms, and thanks to experiments involving giant particle accelerator machines ('atom smashers') (fig. 124), physicists are confident that they understand the physics of nuclear densities thoroughly – and, of course, they understand what goes on at lower densities at least as well. If we wound back the expansion of the Universe to a time when the density everywhere was the same as the density of the nucleus of an atom today, we would go all the way back to 0.0001 of a second (one ten-thousandth of a second) after time zero. Cosmologists are confident that they understand, in general terms, everything that has happened to the Universe since then.

At that time, the temperature of the Universe was 1000 billion degrees (10^{12}) K, and the photons of what would become the background radiation carried so much energy that they were interchangeable with particles. As the Universe expanded and cooled over the next four minutes, some of the photons condensed out as protons and neutrons, the building blocks of atomic nuclei, while the temperature fell steadily. A quarter of these particles ended up as nuclei of helium, and the rest stayed as lone protons, hydrogen nuclei. But it was still too hot for these nuclei to attract electrons and become neutral atoms; for a while, the electrons continued to interact with the radiation.

Fig. 124. One of the underground experimental areas of the Large Hadron Collider (LHC) particle accelerator machine being built at CERN. This cavern will be the home of CMS, a detector designed to search for the Higgs boson.

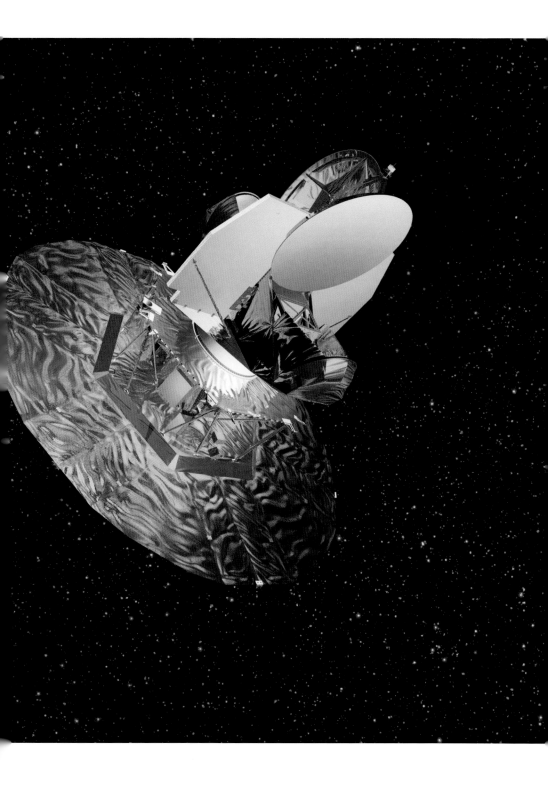

Fig. 125 (overleaf)
An artist's depiction of
WMAP in orbit. Launched
on 30 June 2001, WMAP
maintains a distant orbit
about the second Lagrange
Point, or 'L2', four times
farther than the Moon from
Earth, at a distance of more
than 1.5 million km (0.93
million miles).

It took between 300,000 and 400,000 years for the temperature of the Universe to fall below the temperature at the surface of a star, at which point the photons could decouple from the matter and become the background radiation we detect today. All this time, the atomic (or baryonic) matter was embedded in a sea of dark matter, which does not interact with radiation. Small irregularities in the distribution of the dark matter produced regions of higher density, whose gravity pulled both kinds of matter into clumps, linked together along filaments of dark matter like beads on a string. These clumps attracted more matter by gravity, like water flowing into a pothole in the road.

By about 20 million years after time zero, dark matter potholes were attracting streams of baryonic material, which formed stars and galaxies as it was concentrated by the gravity of the dark matter. Scarcely more than a billion years after time zero, less than one-tenth of the age of the Universe today, as well as many smaller galaxies there were already some proto-galaxies as large as the Milky Way, embedded in haloes of dark matter containing as much as 1000 billion solar masses of material, with quasars or other forms of black hole activity at their hearts. From then on, galaxies grew and evolved through mergers, as we described earlier.

But cosmologists are not content with knowing how the Universe has developed since 0.0001 seconds after time zero. They want to probe back even farther into the past. This is where the educated guesswork comes in, and there is no single theory, or model, which everybody agrees on. Which model you prefer depends to some extent on your personal taste, and as we don't have room to discuss them all we will only describe the one that we like best.

According to that model, the entire Universe that we can see is just part of a bubble within a much larger region of spacetime. This may be literally infinite in both time and space; in order to distinguish it from what we usually mean by the term Universe, it is sometimes called the Cosmos, and the implication is that there may be other bubble universes spread through the Cosmos like the bubbles in a fizzy drink. If the Cosmos is infinite in time, that means it has always existed, always will exist, and there is neither a beginning nor an end of time for us to worry about. If it is infinite in space that means it extends forever in all directions and there are no edges to worry about.

All this makes it relatively simple to describe in mathematical terms, and one feature of that description is that the equations tell us that it is possible for quantum effects to produce tiny bubbles within the spacetime of the Cosmos. Quantum physics is the branch of physics that describes what happens on the scale of atoms and fundamental particles. Since these quantum fluctuations, as they are known, are much smaller than an atom, they might not seem to have much in common with a Universe as large as ours, which has been expanding for nearly 14 billion years and contains hundreds of billions of galaxies. But at the beginning of the 1980s the American theorist Alan Guth realised that there is a way to make one of these quantum fluctuations expand very rapidly indeed for a short time, growing to become the seed of the Big Bang.

The process is called inflation, and it works in exactly the same way as the dark energy that is making the Universe expand faster today, but much more powerfully. The dark energy associated with inflation could take a quantum fluctuation about 100 billion billion (10^{20}) times smaller than a proton and 'inflate' it into a region 10 centimetres across (about the size of a grapefruit) in a tiny fraction of a second.[15] Then, Guth showed, the dark energy is converted into the energy of a hot fireball, expanding so rapidly after this initial push that even though gravity immediately begins to slow it down it would take hundreds of billions of years to halt the expansion.

Inflation comes with a great bonus. During that first split second, what is now our visible Universe would have doubled in size a hundred times. This had the effect of smoothing out any irregularities, in the same way that the wrinkly surface of a prune smoothes out when the prune is placed in water and swells up. Inflation also makes space very, very flat, just as we see in the real Universe. The surface of the Earth already looks pretty flat, even though we know we live on the surface of a sphere. Imagine doubling the diameter of the Earth a hundred times; doubling once makes it twice as big, doubling twice makes it four times as big, doubling three times makes it eight times as big, and so on. Doubling a hundred times makes it 2^{100} times as big. It would be almost impossible for people living on the surface

15__If you want to know how tiny, in terms of seconds you would need to write 32 zeros after the decimal point before writing a 1.

of such a huge sphere to tell that the surface was curved at all. As far as any measurements were concerned, it would be indistinguishable from a flat surface.

Inflation predicts that the Universe should be flat, in the sense we have already described, and that it should contain only tiny irregularities at the time of the Big Bang. These are exactly the kind of tiny irregularities we see in the background radiation, which we know are exactly the right size to account for the existence of galaxies. The stunning implication of inflation theory is that galaxies exist (and therefore, planets and people exist) thanks to quantum fluctuations that were imprinted when the entire Universe we see around us was smaller than an atom.

Inflation is such a successful explanation of the way the Big Bang got started that it is included in almost all of the modern ideas about how the Universe was born, not just the eternal Cosmos model that we particularly like. The big debate in cosmology today really concerns the way inflation got started, but the huge expansion of the Universe during inflation smoothed away so much information about what went before that we may never know exactly how the Universe got started. It may be, though, that inflation is linked to the fate of the Universe, as well as its origin. When dark energy in its current form has done its work, long after the galaxies have receded so far apart that they are invisible to each other, the dark energy will rip apart the structure of matter itself and produce the ultimate void. These are ideal conditions for the kinds of quantum fluctuations that have been linked to the birth of our Universe. It may be that the death of our Universe will lead to the birth of other universes, and that we are just one link in a cosmic chain of universes extending infinitely far into the past and infinitely far into the future. Another idea is that each quantum fluctuation produces a different 'bubble universe' (fig. 126).

That rather overwhelming image makes us seem pretty insignificant on a cosmic scale. But putting speculation aside, coming back down to Earth and concentrating on our own Universe, we find that there are intimate links between life itself and the Universe at large. Everything we have described so far in this book comes together when we come back from speculations about infinity to look at what astronomers can tell us about the relationship between life and the Universe.

Fig. 126. Artist's impression of the Multiverse.

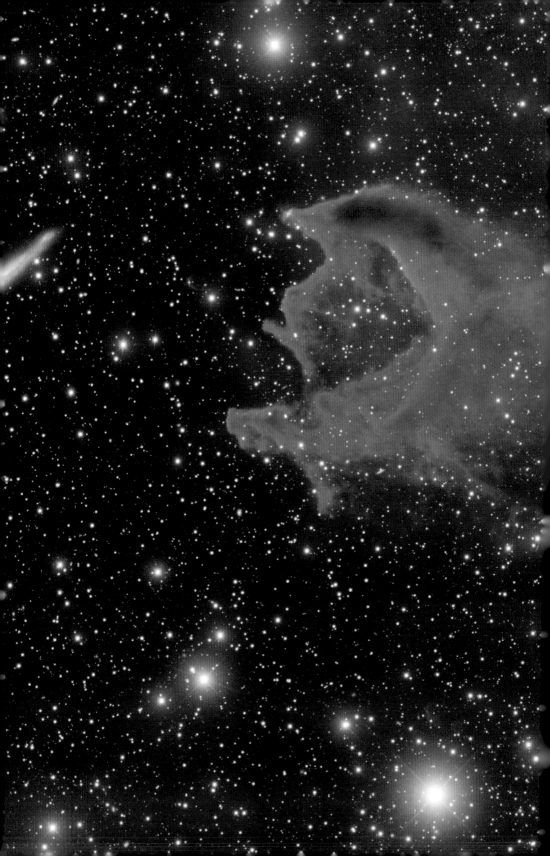

Chapter_Nine
Life and the Universe

The story of life begins with the Big Bang. As far as the kind of matter we are made of is concerned, what came out of the Big Bang was a mixture of very nearly 75 per cent hydrogen and 25 per cent helium, with just a tiny trace of a few other light elements such as lithium. The dark matter was crucial in forming the potholes in which atomic (baryonic) matter could settle and form things like galaxies, stars and planets, while the dark energy will be crucial in determining the fate of the Universe. But neither of them directly affect the story of how clouds of hydrogen and helium gas congregating in gravitational potholes in the expanding Universe got turned into stars, planets and people.

In order to collapse under their own weight and form stars, clouds of gas have to radiate heat away into space. Otherwise the gravitational energy released as heat by the collapse builds up a pressure that stops the collapse. For small clouds, unless the heat can be radiated away the pressure may be so great that it blows the cloud apart before it can form a star. It is very difficult for clouds made almost entirely of a mixture of hydrogen and helium to radiate heat away, so only very massive stars, heavy enough to overcome this pressure, formed when the Universe was young. The first stars contained about 10 to 100 times as much mass as our Sun, and processed hydrogen into helium and helium into carbon and other heavy elements, such as nitrogen, in their cores before they exploded at the end of their lives.

Such big stars have lifetimes of less than a million years, so very soon after the Big Bang the clouds of gas and dust in space were seeded with carbon and other heavy elements from these first stars. This was crucial, because if a hot cloud of gas contains a trace of elements such as carbon the

heavier elements are able to radiate energy away as infrared radiation, which escapes from the cloud and allows it to cool and collapse even if it has less than ten times as much mass as our Sun. The higher the concentration of heavier elements, the easier this is, and the smaller the stars that can be produced.

Second-generation stars with masses of about eight or ten solar masses formed in profusion once the first stars had seeded the clouds in this way, and they dominated the scene for about 30 to 100 million years after the Big Bang. These stars have typical lifetimes of tens of millions of years, so are able to manufacture more elements in their cores before they explode and scatter these products of nuclear burning across the Universe. At last, about 100 million years after the Big Bang, the interstellar clouds were sufficiently enriched with heavy elements to allow stars with only three to seven times as much mass as our Sun to form, and these were the stars that produced the mixture of heavy elements that we find in the Sun and other stars about the same age as the Sun. They dominated the scene from about 100 million years after the Big Bang until a billion or more years after the Big Bang. But the process has not stopped; each generation of stars enriches the interstellar medium more, and stars that are forming in the disc of our Galaxy today contain even more in the way of heavy elements than the Sun does. Older stars, though, as we have seen, retain a memory of what conditions were like when they were born.

Galaxies like the Milky Way, with the two distinct populations of stars that we mentioned earlier, were already in existence by 4 billion years after the Big Bang, about 10 billion years ago. Ever since then, the cycle of star birth, life and death has continued in a more or less steady state in the disc of the Milky Way itself, in the way that we have already explained. The Sun itself, and our Solar System, formed a bit less than 5 billion years ago, when the Milky Way was half its present age.

The clouds from which stars like the Sun formed contain a wide variety of chemical elements. But these are not just in the form of simple atoms. Many of these atoms in interstellar clouds exist in various kinds of molecular compounds. Some of these molecules can be made up of large numbers of different kinds of atoms. They can be detected and studied from Earth because each kind of molecule radiates or absorbs energy in a particular part of the radio band of the electromagnetic spectrum, in exactly the same way that atoms can be identified from their characteristic lines in the spectrum of visible light.

In fact, there are almost too many of these radio fingerprints. The problem is not finding the lines in the radio part of the spectrum but matching them up with the appropriate molecules. The only way to do this is to study the radiation from molecules in the laboratory, and in a kind of Catch-22 situation you have to know which molecules to make in order to identify their spectra! Partly by trial and error, partly by luck and partly by inspired guesswork astronomers have now identified the radio fingerprints of well over a hundred molecules in space. Hardly surprisingly, the first to be identified were some of the simplest – including water vapour (H_2O) and ammonia (NH_3). But the study of molecules in space really took off in 1969, with the discovery of formaldehyde, which has the chemical formula H_2CO, in interstellar clouds.

Formaldehyde set astronomers thinking for several reasons. It contains carbon, and complex molecules that contain carbon are known as organic compounds, because they are very important for life. Specifically, formaldehyde itself is a key component of life, as it is one of the building blocks of more complicated organic molecules such as sugars, which are directly involved in life processes (DNA is largely composed of sugar molecules). So the discovery of formaldehyde in space gave astronomers the incentive for the tedious work needed to identify many more of the radio fingerprints they could already detect, as well as to search for more signs of molecules in space.

The molecules identified so far include long (but rather boring) chains in which up to eleven carbon atoms are linked in a row, with a hydrogen atom at one end and a nitrogen atom at the other, many organic compounds familiar to terrestrial chemists (such as ethyl alcohol, formic acid and hydrogen cyanide) and very large molecules made up of rings of carbon atoms joined together (*fig. 129*). Each of these rings is made of six carbon atoms, little hexagons that can attach to other atoms or molecules at their corners. They can also join together with other carbon hexagons, making up sheets known as polycyclic aromatic hydrocarbons

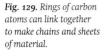

Fig. 129. *Rings of carbon atoms can link together to make chains and sheets of material.*

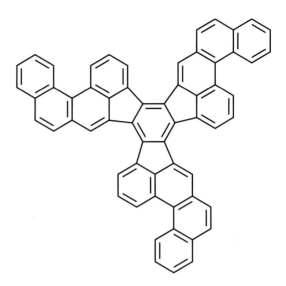

(PAH). One such sheet can contain a hundred or more carbon atoms, with hydrogen atoms and perhaps a few other atoms attached round the edges of the sheet.

Without going into details of their names and specific properties, the complexity and variety of some of the molecules found in space so far can be seen from their chemical formulae – things like SiO, SO_2, C_2H_2, C_3H_2, H_2HCO, $HCOOH$, CH_3HCO, CH_3CN, $HCONH_2$ and CH_2OHCHO. The last one on this list, glycoaldehyde, is a sugar that is found in large quantities in interstellar clouds, and adding two more hydrogen atoms to glycoaldehyde makes ethylene glycol, made of two linked CH_2OH units, a ten-atom molecule that is also found in space. It is one of the largest found so far (apart from PAH sheets) and is the active ingredient of antifreeze. Experiments in laboratories on Earth show that glycoaldehyde can react with another simple sugar to produce ribose. Ribose is the key component of ribonucleic acid, or RNA, and if oxygen is removed from RNA molecules you get deoxyribonucleic acid (DNA). Some of the other molecules in interstellar clouds, notably the ones containing nitrogen as well as hydrogen and carbon, are the kind of raw material from which amino acids, the structural units of protein molecules, are made. Astronomers have not yet found the building blocks of life in space. But they have found the building blocks of the building blocks of life.

These complex molecules are able to grow in space because of the presence of tiny grains of graphite (carbon) and silicate (an oxide of silicon) in the material ejected from old stars. These particles are able to form because they go straight from the gaseous state to a solid state as the material around the star cools, without passing through a liquid phase. The grains are tiny, like the particles in cigarette smoke, but they soon become embedded in an icy shell of material – not just frozen water, but other kinds of ice including frozen methane, frozen ammonia and frozen carbon monoxide.

If any large molecules were floating freely in space and then collided with another molecule, the chances are that the impact of the collision would break the molecules apart, not cause them to stick together to make an even bigger molecule. But the frozen grains in interstellar clouds provide surfaces where molecules can stick, and where another molecule colliding with the icy surface will have its impact energy taken up by the grain instead of disrupting the molecules that are already there. The last ingredient needed to encourage the chemical reactions that build up large molecules is a source of energy, and this comes from ultraviolet radiation from the stars. All these ideas have been tested in chemical laboratories, where icy grains cooled to 10 K have been fed with a suitable chemical mixture and dosed with ultraviolet light.

Overall, across the Milky Way interstellar space contains about one-tenth as much baryonic material as all of the bright stars in the Galaxy put together. This adds up to a few tens of billion times the mass of our Sun. For every new star that is born out of this mixture, very nearly as much mass is added to it as old stars die, so the interstellar matter is constantly being enriched and recycled, but it is very slowly getting used up as some stars end their lives as white dwarf stars, neutron stars or black holes.

When a planetary system like the Solar System forms, planets like the Earth are, as we have seen, born in fire, and pass through a completely molten stage. There is no chance of any complex organic molecules surviving under such conditions. So until recently it used to be thought that when the Earth was young the energy available from lightning bolts might have triggered chemical reactions that started out from simple compounds like water, methane, carbon dioxide and ammonia and produced the building blocks of life, such as amino acids. Experiments in which such mixtures have been sealed in glass flasks and subjected to electric discharges suggest that it might work, but it would be a very slow process. And yet, the fossil evidence tells us that single-celled life appeared on the surface of the Earth very soon after the planet cooled to the point where liquid water could flow, nearly 4 billion years ago.

If you asked a chemist to synthesise amino acids in the lab, he or she would not start out with things like methane and ammonia, but with more complex chemical compounds such as formaldehyde, methanol and formamide. With these ingredients, manufacturing amino acids is quick and easy. All of these compounds are known to exist in the kind of clouds from which stars and planetary systems form, and astronomers are confident that before too long they will identify amino acids themselves in space. After all, many complex organic molecules including amino acids have been found in meteorites, pieces of material left over from the formation of the Solar System.

The key to the origin of life on Earth comes from the evidence that there was a lot of liquid water and life on our planet 3.8 billion years ago, less than a billion years after the Earth formed. Where did all the water come from? The answer is comets. A great deal of the material in the cloud of stuff that collapsed to form the Solar System was in the form of icy lumps, and although any icy lumps in the inner part of the Solar System where the Earth formed would have been evaporated by the heat of the young Sun, thousands of billions of them remained intact orbiting in the cooler outer regions of the young Solar System. When the giant planets formed, their gravitational influence sent vast numbers of these comets hurtling inwards (and deflected just as many outward), where some of them collided with the inner planets. Various dating techniques tell us, especially when applied to the battered face of the Moon, that the resulting bombardment lasted for about half a billion years, and ended just about 4 billion years ago.

Some idea of just how powerful the impact of a comet with a planet can be was seen in 1994, when pieces of ice about 10 km (6 miles) across, fragments of a large comet broken up by tidal forces in Jupiter's gravitational grip, struck the giant planet, producing scars that pock-marked its surface (see *fig. 70, page 114*). The impact of an object about the size of one of these fragments with the Earth is also thought to have been responsible for the death of the dinosaurs, some 65 million years ago.

Comets that struck the face of the Moon during the ancient bombardment left large craters, but the gravity of the Moon is so feeble that the volatile materials that formed the bulk of the comets easily escaped back into space. But most of the gas that forms the Earth's atmosphere, as well as the water to form the Earth's oceans, arrived in this way and stayed, thanks to the Earth's stronger gravitational pull. Complex organic molecules might not have survived the extreme conditions of these impacts, in which energy equivalent to the explosion of hundreds of millions of megatonnes of TNT was released as heat, but once the

Fig. 130. *A bright meteor streaks horizontally through the sky during the annual Perseid meteor shower, viewed from Rabbit Ears Pass, Colorado in 2001.*

Earth had an atmosphere it provided a buffer that allowed a lot of material to fall to the surface more slowly. Debris from comets that have broken up in space still falls to Earth today. Some of this is responsible for showers of meteors, such as the Perseids seen every August (*figs. 130, 131*). These 'shooting stars' are caused by the burn up of sand-grain or pea sized particles in the upper atmosphere. But smaller particles of dust from comets also drift into the atmosphere more gently and float down to the surface of the planet. These tiny grains have been monitored by satellites and trapped using high flying aircraft and brought back down to the ground to study.

Fig 131. Artist's impression of a Perseid meteor shower.

Fig. 132. Earth from space: this true-colour satellite image shows a large phytoplankton bloom off the west coast of Tasmania.

Compared with the size of the Earth, there is not a lot of this material reaching the ground today. But by human standards it is impressive – even today, about three hundred tonnes of organic material reaches the surface of the Earth from space in this way each year, along with much more inorganic material. At the end of the primordial battering of the Earth's surface, after the atmosphere and oceans had formed, about ten thousand tonnes of organic material reached the surface in this way each year. In three hundred thousand years, that would have added up to the equivalent of the mass of all the living things on Earth today. That was the raw material, energised by lightning bolts and ultraviolet radiation, from which life emerged before another 200 million years had passed.

Just how life emerged remains a mystery, but it is clearly a lot easier to do the trick starting from such a rich brew of organic compounds than starting from water, methane, carbon dioxide and ammonia. Some astronomers even speculate that life itself might have appeared first in space, in the form of single-celled organisms living in comets, where there was energy from starlight and perhaps from radioactivity to encourage chemical reactions. The chemical processes would have been slow, but there would have been nearly 10 billion years available from the time of the Big Bang to the formation of the Solar System.

We may never know if this speculation is correct. But we do know that, like everything else on Earth except for hydrogen and helium, living things, including ourselves, are recycled stellar debris. All the chemical elements except for hydrogen and helium (and even some of the helium) have been manufactured inside stars and thrown out into space to provide the raw material of later generations of stars, planets, and in at least one case people. We are, indeed, made of stardust. We also know from this that any Earth-like planet existing in a galaxy like the Milky Way will, like the Earth, be seeded with complex organic molecules that are the precursors to life. If Earth-like planets are common, life ought to be common. And planets now seem to be very common indeed.

The easiest way to find a planet orbiting another star is by observing the way the gravity of the planet tugs on the parent star, pulling it from side to side as the planet moves around the star. This produces a tiny but regularly changing Doppler shift in the spectrum of light from the star, and astronomers are very good at measuring Doppler shifts. The snag is that the effect is far too small to measure for a planet like the Earth orbiting a star like the Sun. It works best for large planets in orbits close to their parent stars. So, hardly surprisingly, once astronomical detectors were sensitive enough to measure such effects, in the middle of the 1990s, these were the first kinds of extrasolar planets (see *fig. 136*, pages 226–7) to be discovered.

The first of these Jupiter-like planets was found orbiting a star called 51 Pegasi, a star similar to the Sun, some 50 light years away. The planet has about half the mass of Jupiter, but orbits the star so closely that its year is just four of our days, making it relatively easy to spot the tell-tale

Doppler wiggle in the spectrum of light from the star. Before long many other 'jupiters' orbiting other stars were discovered, most with masses several times that of Jupiter itself, and all in orbits that result in a pronounced Doppler effect, making them easy to detect. The discovery that many of these planets are in orbits much closer to their parent stars than the present orbit of Jupiter around the Sun reinforces the idea that large planets form close in to stars and migrate outwards as time passes. But it does not mean that most planetary systems today are like the ones we have found so far. Astronomers suspect that there must be many planetary systems with Earth-like planets in Earth-like orbits that we simply cannot detect yet. So far, we have been able to search for planets out to a distance of only 150 light years, a tiny region of the Milky Way, and six per cent of the stars studied have been found to be accompanied by large planets. A couple of hundred extrasolar planets have now been discovered, with more being added to the list each year.

In principle, the Doppler technique could be used to find Earth-like planets in Earth-like orbits, but it would be very difficult. Jupiter itself causes a wobble of the Sun that produces a velocity shift of just 12.5 metres (41 feet) per second, which is a little faster than the speed of the world 100-metres record. It would be possible to measure a shift that small in the light from a nearby star, but Jupiter takes just over 11 years to go round the Sun once, and astronomers need two orbits' worth of data to be sure of measuring the effect. So it would take more than 20 years of observations to find a single planet like Jupiter in an orbit like Jupiter's orbit around a nearby star. It would only take two years to find a planet in an orbit like that of the Earth, if the Doppler shift were big enough. But the Earth's influence on the Sun is only 1 metre (3 feet) per second, equivalent to a very gentle walking pace. It is too small to be detectable using present-day techniques.

Fortunately, though, there is another way to detect Earth-like planets in Earth-like orbits. If the orbit of a planet going round another star happens to be lined up in the right way, the light from the star will be dimmed slightly, in a kind of mini-eclipse, when the planet passes in front of it. These transits, as they are called, will be rare, since they will occur only once for each orbit of the planet around its star, and they are also short-lived. But there are plans to launch satellites to monitor large numbers of stars to search for such effects. The way the statistics work out, if 100,000 stars were studied in this way and 1000 of them (just 1 per cent) showed the right kind of dimming, that would imply that almost every Sun-like star has a family of planets similar to the inner planets of the Solar System.

Once astronomers do find Earth-sized planets, they know how to look for signs of life. It all comes back to Jim Lovelock's idea of Gaia, which sees the living Earth as a self-regulating system. Lovelock realised that it is the presence of life on Earth that, among other things, maintains an atmosphere rich in reactive substances like oxygen and water vapour, which ought to react chemically to make inert substances. The discovery that Mars has an inert atmosphere of carbon dioxide is evidence that there is no life on Mars, even without going there to look for life. But, by the same token, if you could detect oxygen, water vapour and other reactive substances in the atmosphere of a distant planet, you would know that there *is* life there, without going to the planet to look for it.

Anticipating the imminent discovery of other earths, both the European Space Agency (ESA) and the American Space Agency (NASA) are already working on plans for a

Fig. 133. *The spectrum of the Earth seen from space. This shows a pronounced dip near 10 micrometers due to the presence of ozone. This is one of the key signatures of life.*

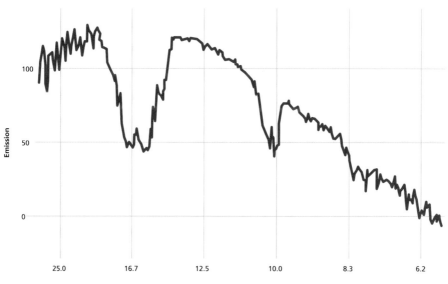

Emission

Wavelength (mμ metres)

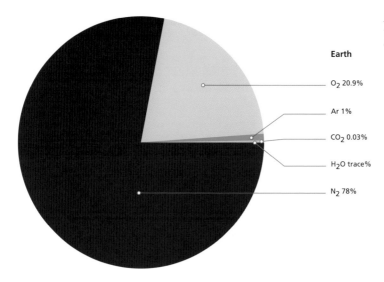

Fig. 134. *The atmospheric make-up of Earth, Venus and Mars.*

Earth

O_2 20.9%

Ar 1%

CO_2 0.03%

H_2O trace%

N_2 78%

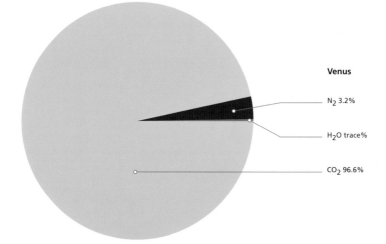

Venus

N_2 3.2%

H_2O trace%

CO_2 96.6%

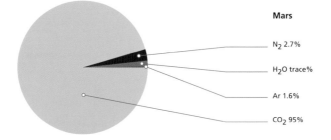

Mars

N_2 2.7%

H_2O trace%

Ar 1.6%

CO_2 95%

space telescope that will be able to make such observations. Such a telescope will be designed to be sensitive
to radiation in the infrared part of the spectrum, because
planets like the Earth radiate strongly in the infrared (*fig.
133*) – they absorb energy from their parent stars and re-
radiate it as infrared heat. We know what the spectrum of
the Earth itself is like, both from theoretical calculations
and by looking at our planet from outside using spacecraft.
There are three pronounced dips in the spectrum, corresponding to absorption by carbon dioxide, water vapour,
and the highly reactive form of oxygen called ozone. The
detection of carbon dioxide alone would be the easiest
task, and would show that the planet being studied at least
had an atmosphere, like Venus or Mars. The detection of
water vapour would imply that it had lakes and oceans,
which are essential for life on Earth. And the presence of
ozone, the hardest of the three to detect, would be evidence
that there is a large amount of oxygen in the atmosphere,
so there must be life on the planet. (See *figs. 134, 136*).

In order to make the necessary observations, the telescope would have to be placed far out in space near the orbit
of Jupiter, away from any sources of infrared interference
in the inner Solar System. It would probably consist of six
infrared telescopes, each in its own unmanned spacecraft,
flying in formation at the corners of a hexagon, with a central spacecraft collecting data from the six telescopes and
relaying it back to Earth. By flying in perfect formation,
kept in place by using laser beams linking all the spacecraft
to monitor the separation between them, this arrangement
would mimic the power of a much larger single telescope.
The separation between individual spacecraft (the length of
the sides of the hexagon) would be about 100 metres (328
feet), but their position would have to be maintained to an
accuracy of less than 1 mm (0.04 in), all while flying in space
more than 600 million km (373 million miles) from Earth.
The astronomers and space scientists are, however, confident that they have the technology to do this by the year
2020, if they are given the financial backing needed to do
the job.

Such a telescope will be able to obtain images of Earth-
sized planets, if they exist, around a few hundred stars
within about 50 light years of the Solar System. The astronomers already have a target list of 200 good candidates for

Fig. 135. *From back to front: Earth, Venus and Mars, shown to scale.*

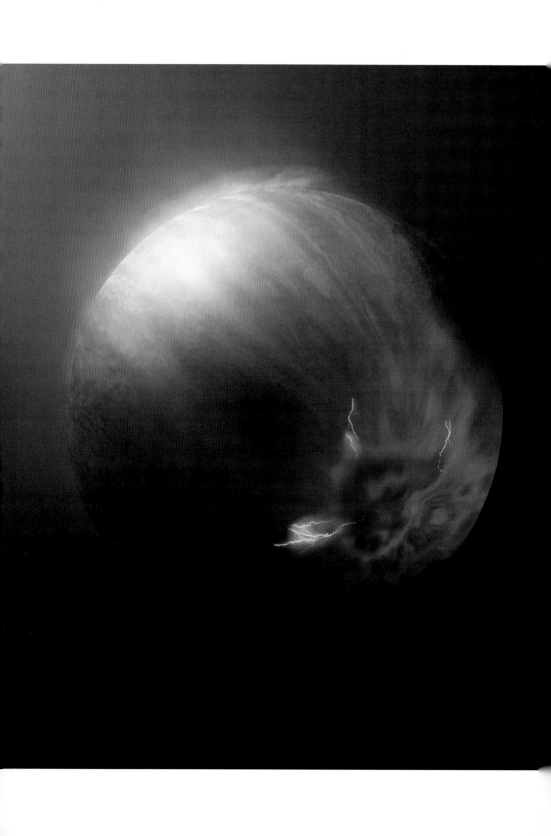

Fig. 136 (overleaf)
There are almost 300 extrasolar planets known to exist. Here are some imaginative examples – large planets where global winds are whipped up by heat generated in the planet's atmosphere by its nearby primary star – while huge lightning bolts discharge in the dark as this energy is released on an unimaginable scale.

such a study. For each star, an observation lasting a few tens of hours would be enough to get an image of planets equivalent to Venus, Earth and Mars; planets like Mercury would be lost in the glare of their parent star. Making these preliminary observations would take a few thousand hours of telescope time in all, the first year or so of the telescope's operation. The images would be no more than fuzzy blobs, but repeated observations over the course of months or years would show them moving in their orbits around their parent stars.

Obtaining the basic spectra of the planets and searching for the features produced by carbon dioxide and water vapour would take about 200 hours for each planetary system being observed. For a spacecraft with a lifetime in

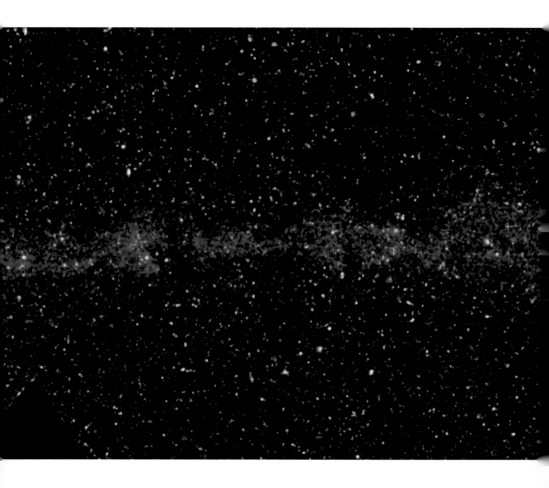

space of about five years, this means that there would be time to make such observations of about 80 planetary systems, over the next couple of years, leaving nearly half the lifetime of the mission to concentrate on the search for ozone. This would require about 800 hours of observation for each planet, giving time to analyse the spectra of the 20 or so most promising candidates. So it is possible that by the year 2025 we will know for sure that there are up to 20 Earth-like planets with oxygen-rich atmospheres that are homes for life beyond the Solar System. But even if only one such planet is found, it will be proof that the Earth is not unique, and that life is a natural feature of the Universe. It would be the most significant discovery ever made by humankind.

Fig. 137. One of the strangest objects in the sky, the old red giant star Mira is hurtling through space at 130 km (81 miles) per second, so fast that it is shedding material behind it, making a tail like that of a comet but 13 light years long. As Mira speeds along, the debris it leaves in its wake contains carbon, oxygen and other elements needed for new stars, planets and life to form. The material making up the tail has been released over the past 30,000 years.

Glossary

Absolute magnitude A measure of the actual brightness (*magnitude*) of a *star*, defined in terms of the brightness it would have if viewed from a distance of 10 *parsecs*. The absolute magnitude of the *Sun* is 5.

Astronomical unit (AU) A measure of distance used by astronomers, equivalent to the average distance between the *Earth* and the *Sun*, 149,597,870 km (92,955,810 miles). Particularly useful for measuring distances across the *Solar System*.

Atom The smallest component of an element of everyday matter such as oxygen or carbon. Elements combine to make compounds in which the atoms are joined together to make molecules such as carbon dioxide. An atom has a tiny central nucleus, with a positive electric charge, surrounded by a cloud of *electrons* that have a negative charge that exactly balances the positive charge on the nucleus. The size of the nucleus compared with the size of the electron cloud is like the size of a grain of sand compared with the size of the Albert Hall; but almost all of the *mass* of the atom is in the nucleus. It would take 10 million atoms side by side to reach across the space between two of the points on the serrated edge of a postage stamp.

Baryonic matter Name given to matter like the everyday matter here on *Earth*, made of *protons*, *neutrons* and *electrons*.

Big Bang Theory The scientific description of how the Universe as we know it emerged from a hot, dense fireball nearly 14 billion years ago.

Black hole A region of space where the gravitational field is so intense that nothing, not even light, can escape. Some black holes contain only a few times as much *mass* as our *Sun*; others contain millions of times as much mass.

Blueshift See *Doppler effect*.

Cepheid A kind of variable *star* that changes brightness in a regular way that enables astronomers to work out its average brightness and therefore how far away it is.

Comet A lump of icy debris in the *Solar System*, typically tens or hundreds of kilometres across. When a comet moves close to the *Sun*, gas streaming off from it produces a bright 'tail' giving it a distinctive appearance.

Cosmic Background Radiation (CBR) A weak hiss of radio noise coming from all directions in space, the echo of the *Big Bang*.

Cosmological redshift The stretching of light from a distant object to longer wavelengths caused by the expansion of the Universe. (This is not the same as the *Doppler effect*.)

Dark energy A form of energy that fills all of space, detected only by its influence on the way the Universe expands.

Dark matter Material detected only by its gravitational pull that affects the way galaxies move and how the Universe expands. There is seven times more dark matter than there is *baryonic matter*.

Doppler effect A change in the wavelength of light or sound that happens when the object emitting the waves is moving towards or away from the observer. Light from an object moving towards you is squashed to shorter wavelengths (*blueshift*); light from an object moving away from you is stretched to longer wavelengths (*redshift*). See also *Cosmological redshift*.

Dwarf planet A small ball of rock or gas, big enough for *gravity* to make it round, in orbit around a *star*. See *planet*. Just where you draw the line between planets and dwarf planets is not clearly defined, but astronomers know a dwarf planet when they see one.

Earth The *planet* we live on.

Earth System Science See *Gaia theory*.

Electromagnetic radiation A form of radiation in which energy is carried by oscillations of combined electric and magnetic fields moving through space. Light is a form of electromagnetic radiation, and all electromagnetic radiation travels at the speed of light. Moving packets of electromagnetic radiation can also be described, using quantum theory, in terms of *photons*, the 'particles' of light.

Electron The lightest of the building blocks of matter, an electrically negative particle found in the outer part of an *atom*.

Element See *Atom*.

Ethane A compound made of *atoms* of hydrogen and carbon; found in the atmospheres of some of the giant *planets*.

Equinoxes The two days in the year, 21 March and 23 September, when there are exactly 12 hours of daylight and 12 hours of night everywhere on *Earth*.

Extrasolar planet Any *planet* orbiting a *star* that is not the *Sun*.

Gamma ray A very energetic *photon*.

Gaia theory The description of all the components of the *Earth*, including both living things and non-living things such as volcanoes, as a single system of interacting units. Also known as *Earth System Science*.

Galaxy A large island in space containing many *stars* like the *Sun* – up to hundreds of billions of them.

Globular cluster A densely packed ball of *stars* that may contain millions of individual stars.

Gravity A force exerted by any object with *mass* on any other object with mass. The size of the force depends on the mass of the object divided by the square of the distance from it. This is called the inverse-square law of gravity.

Helioseismology Technique for probing the *Sun's* interior by studying waves on its surface. Equivalent to using earthquakes to probe the interior of the *Earth* (seismology).

Hubble constant Also known as the Hubble parameter, a number that measures how fast the Universe is expanding.

Infrared A form of invisible light with wavelengths longer than red light. See *Spectrum*.

Kelvin temperature scale Temperature measured from the absolute zero of temperature, − 273.15 °C. Each degree on the Kelvin scale is the same size as a degree on the Celsius scale, but is written without the 'degrees' sign. So 0 °C is the same as 273.15 K, and so on.

Late heavy bombardment (LHB) The battering received by the *Earth*, *Moon* and inner *planets* of the *Solar System* a little more than 4 billion years ago as they swept up the debris left over from the formation of the planets. The LHB was responsible for the heavily cratered surface of the Moon.

Light year A measure of distance used by astronomers, equivalent to the distance light can travel in a year, 9.46 million million km (5.9 million million miles).

Magnetosphere The zone around a *planet* in which its magnetic field dominates over the magnetic influence of its parent *star*.

Magnitude A measure of the brightness of astronomical objects as seen from *Earth*. For historical reasons, a difference in magnitudes of 5 means that one object is 100 times brighter (or fainter) than another. A difference of one magnitude therefore corresponds to a difference of just over 2.5 times in brightness. See also *Absolute magnitude*.

Main sequence star A *star* in the quiet prime of its life, like the *Sun*.

Mass A measure of the amount of matter something contains. It is mass that causes an object's resistance to being pushed around (inertia), and it is mass that produces the gravitational force that attracts objects to one another. The bigger the mass, the bigger the inertia and the bigger the gravitational pull.

Meteor The streak of light across the sky caused when a piece of cosmic debris burns up in the *Earth's* atmosphere.

Meteorite Strictly speaking, the name for a lump of material from space that strikes the surface of the *Earth*. By extension, used to refer to objects striking the *Moon* or the inner *planets* of the *Solar System*.

Methane A compound made of *atoms* of hydrogen and carbon; found in the atmospheres of some of the giant *planets*.

Milky Way The name of our home *galaxy*; also called the Galaxy, with a capital 'G'.

Molecule See *Atom*.

Moon With a small 'm', any natural object in orbit around a *planet*. Moons may be small or large, rocky or icy, irregular or round, with or without an atmosphere. 'The' Moon, with a capital M, is the name of the *Earth's* moon.

Nebula A cloud of gas and dust between the *stars*. Because *galaxies* were once thought to be part of the *Milky Way*, some of them, like the Andromeda Galaxy, were labelled nebulae; but as applied to galaxies the term is now obsolete.

Neutrino A very light particle produced in nuclear reactions. Many neutrinos escape from the inside of the *Sun* and give astronomers insight into what goes on there.

Neutron One of the building blocks of matter, an electrically neutral particle found in the nucleus of an *atom*.

Neutron star An extremely compact stellar remnant containing about as much *mass* as our *Sun* in a sphere no more than 10 km (6.2 miles) across.

Nucleosynthesis The natural processes that build up heavier *elements* from lighter ones. A little of this happened in the *Big Bang* (Big Bang nucleosynthesis) but most of the elements except hydrogen and helium have been manufactured inside *stars* (stellar nucleosynthesis).

Nucleus See *Atom*.

Ozone A form of oxygen in which there are three *atoms* in every molecule, instead of the two there are in the air we breathe. Because of the action of sunlight on oxygen, ozone forms in a layer of the atmosphere called the stratosphere, high above the ground. If telescopes detect ozone in the *spectrum* of light from a *planet* orbiting another *star*, that will be proof that the atmosphere of that planet contains oxygen, and that it is a home for life.

Parsec A measure of distance used by astronomers, equivalent to 3.2616 *light years*, or 30.85 million million km (19.17 million million miles).

Photon A particle of light or other forms of *electromagnetic radiation*.

Planet A large ball of rock or gas, big enough for *gravity* to make it round, which is in orbit around a *star*.

Planetisimal A piece of rock or ice, a kilometre or so across, which is one of the building blocks from which *planets* are made.

Population I Young *stars* like our *Sun*, relatively rich in heavy *elements*.

Population II Old *stars* that do not contain as many heavy *elements* as *Population I* stars.

Precession A slow wobble of the *Earth* and other *planets* on their axes, like the wobble of a spinning top that is not quite vertical. In the case of the Earth, this means that the direction the North Pole points to on the sky traces out a circle once every 26,000 years.

Proton One of the building blocks of matter, an electrically positive particle found in the nucleus of an *atom*.

Pulsar A very small *star* at the end of its life, a *neutron star* no bigger than a large mountain on *Earth*, but containing roughly as much *mass* as our *Sun* and spinning very fast. Pulsars are easy to find because they emit beams of radio noise that sweep around as it spins, like a natural radio 'lighthouse' in space.

Quasar Extreme example of an active *galaxy*, powered by the energy released as matter is swallowed up by a *black hole*.

Red giant A *star* in the later stages of its evolution, when it swells up to have a diameter about as large as the diameter of the *Earth's* orbit today.

Redshift See *Cosmological redshift*; *Doppler effect*.

Resonance When two varying effects move in step and amplify each other. The famous example of an opera singer shattering a wine glass by singing a pure note is resonance – the musical note matches the natural vibration of the glass.

Satellite Another name for a *moon*; also used to refer to spaceprobes in orbit around a *planet*.

Solar mass The *mass* of our *Sun*, 1.9891×10^{30} kilograms. This is the standard unit used to express masses of *stars*.

Solar System Our *Sun* and its family of *planets*, *moons* and cosmic debris.

Solstices The two days in the year mid-way between the equinoxes. They occur on 22 June and 22 December at present, although these dates change slowly over thousands of years.

Spacetime The unification of space and time in one package, described by the equations of the special and general theories of relativity.

Spectroscopy Technique for analysing what things are made of by studying their light. See *spectrum*.

Spectrum The rainbow pattern of coloured light seen when white light is split up using a prism. The range of colours we can see extends from red through orange, yellow, green, blue and indigo to violet. Red has the longest wavelength, violet the shortest.

Star A hot ball of gas, many times bigger than a *planet*, which shines because energy is released by nuclear reactions going on in its interior. The *Sun* is a star.

Stellar nucleosynthesis See *nucleosynthesis*.

Sun The *star* at the centre of our *Solar System*.

Troposphere The lowest layer of the *Earth's* atmosphere, from the ground up to an altitude of about 15 km (9.3 miles).

Ultraviolet A form of invisible light with wavelengths shorter than violet light. See *Spectrum*.

White dwarf A compact stellar remnant containing about as much *mass* as our *Sun* in a sphere about the size of the *Earth*.

x-ray An energetic *photon*.

Further Reading

Universe, Dorling Kindersley, London, 2007.

Jim Al-Khalili, *Black Holes, Wormholes and Time Machines*, IOP, London, 1999.

A. Bevan, *Meteorites*, Smithsonian, Washington, 2002.

Robert Burnham, *Great Comets*, Cambridge UP, 2000.

Eric Chaisson & Steve McMillan, *Astronomy Today vol 1: The Solar System*, Cummings, New York, 2007.

Eric Chaisson & Steve McMillan, *Astronomy Today vol 2: Stars and Galaxies*, Cummings, New York, 2007.

Marcus Chown, *The Magic Furnace*, Vintage, London, 2000.

Stuart Clark, *Deep Space*, Quercus, London, 2008.

Paul Davies, *The Goldilocks Enigma*, Penguin, London, 2007.

Simon Goodwin, *Hubble's Universe*, Constable, London, 1996.

John Gribbin, *Companion to the Cosmos*, Weidenfeld & Nicolson, London, 1996.

John Gribbin and Mary Gribbin, *Stardust*, Allen Lane, London, 2000.

Stephen Hawking, *The Universe in a Nutshell*, Bantam, London, 2001.

James Lovelock, *The Ages of Gaia*, Oxford UP, 1995.

Stephen Marshak, *Earth*, Norton, London, 2005.

Iain Nicolson, *Dark Side of the Universe*, Canopus, London, 2007.

Ian Ridpath, *Atlas of Stars and Planets*, Philip's, London, 2004.

Carl Sagan, *Cosmos*, Abacus, London, 1983.

Giles Sparrow, *The Planets*, Quercus, London, 2006.

David Whitehouse, *The Sun*, Wiley, London, 2005.

Index

Acknowledgements

We thank the University of Sussex for providing us with a base from which to work, and the Alfred C. Munger Foundation for financial support.

Picture Acknowledgements

Illustrations are reproduced by kind permission of the following individuals or institutions:

akg-images, London: *fig. 119*
Alcatel-Lucent/Bell Labs: *fig. 120*
BBSO/NJIT (photo © Thomas Spirok, 2006): *fig. 53*
British Library, London: *fig. 6*
The Carnegie Institution of Washington (courtesy of the
    Observatories): *figs 90, 106*
EFN – Environmentalists for Nuclear Energy (photo © Bruno
    Comby): *fig. 16*
Fred Espenak: *fig. 47*
HST Comet Team/NASA/ESA: *fig. 70*
Huntington Library, San Marino, California: *fig. 115*
Institute and Museum of the History of Science, Florence: *fig. 4*
Jimmy Westlake, Colorado: *figs. 83, 130*
Kamioka Observatory, ICRR (Institute for Cosmic Ray Research),
    University of Tokyo: *fig. 49*
David Malin Images: *figs 81, 129* (photo © Akira Fujii/DMI); *88* (photo
    © Anglo-Australian Observatory/DMI); *84* (photo © Anglo-
    Australian Observatory/Royal Observatory, Edinburgh/DMI); *37*
    (photo © David Miller/DMI)
NASA: *figs 12, 15, 19, 26, 31, 61, 62; 11* (Goddard Space Flight Center/
    Reto Stöckli/Robert Simmon); *28* (GRIN); *32, 34* (GSFC); *113*
    (Jeff Hester and Paul Scowen (Arizona State University)); *68* (Johns
    Hopkins University Applied Physics Laboratory/Southwest
    Research Institute); *14* (JSC); *102* (NOAO, ESA and The Hubble
    Heritage Team STScI/AURA); *58* (STScI); *125* (WMAP); *94*
    (x-ray: NASA/CXC/Rutgers/J.Warren et al.; Optical: NASA/STScI/
    U. Ill/Y.Chu; Radio: ATCA/U. Ill/J.Dickel et al)
NASA/European Space Agency (ESA): *figs 103, 105; 91* (Andrew
    Fruchter and ERO team (STScI and ST-ECF)); *65, 89, 96* (C.R.
    O'Dell/Rice University); *75* (E. Karkoschka (University of Arizona));
    67 (G. Bacon (STScI)); *74* (J. Clarke (Boston Univ.)/Z. Levay (STScI));
    frontispiece and *114* (S. Beckwith (STScI) and the HUDF Team)
NASA/ESA/Hubble Heritage Team/STScI/AURA: *figs 107, 108, 112;*
    104 (Robert A. Fesen (Dartmouth College, USA)/James Long
    (ESA/Hubble)); *109* (J. Gallagher (University of Wisconsin),
    M. Mountain (STScI) and P. Puxley (NSF))

NASA and Jet Propulsion Laboratory (JPL): *figs 18, 29, 39, 54, 55, 57, 59, 63, 69, 73, 79, 97; 137* (JPL-Caltech); *111* (JPL-Caltech/SSC); *66* (JPL-Caltech/University of Arizona); *100* (JPL-Caltech/University of Virginia); *56* (Massachusetts Institute of Technology, Cambridge, MA/USGS/Flagstaff, AZ); *77* (Space Science Institute); *76, 78* (USGS)
National Maritime Museum, London: *fig. 3*
Nordic Optical Telescope/Romano Corradi (Isaac Newton Group of Telescopes, Spain): *fig. 87*
M. Robberto (Space Telescope Science Institute/ESA) and Hubble Space Telescope Orion Treasury Project Team: *fig. 101*
SeaWiFS Project, NASA/Goddard Space Flight Center/ORBIMAGE: *fig. 132*
Science & Society Picture Library (SSPL): *figs 24* (photo © Jamie Cooper); *40, 50* (NASA); *5, 118* (Science Museum, London)
Sipa Press/Rex Features: *fig. 124*
Greg Smye-Rumsby: *figs 7, 8, 9, 13, 17, 20, 21, 22, 30, 33, 35, 38, 39, 41, 43, 46, 48, 51, 52, 60, 64, 72, 80, 82, 85, 86, 92, 93, 95, 98, 99, 110, 116, 121, 122, 123, 126, 128, 131, 133, 134, 135, 136*
SSDS/Astrophysical Research Consortium (ARC)): *fig. 117*
T.A. Rector/University of Alaska Anchorage/T. Abbott and NOAO/AURA/NSF: *fig. 127*
US Geological Survey (USGS): *figs 10*; (J.D. Griggs) *23, 25*; (Flagstaff) *27*